Machine Learning and Music Generation

Computational approaches to music composition and style imitation have engaged musicians, music scholars, and computer scientists since the early days of computing. Music generation research has generally employed one of two strategies: knowledge-based methods that model style through explicitly formalized rules, and data mining methods that apply machine learning to induce statistical models of musical style. The five chapters in this book illustrate the range of tasks and design choices in current music generation research, applying machine learning techniques and highlighting recurring research issues such as training data, music representation, candidate generation, and evaluation. The contributions focus on different aspects of modeling and generating music, including melody, chord sequences, ornamentation, and dynamics. Models are induced from audio data or symbolic data.

This book was originally published as a special issue of the *Journal of Mathematics and Music*.

José M. Iñesta is a Professor in the Department of Software and Computing Systems at the University of Alicante, Spain.

Darrell Conklin is an Ikerbasque Research Professor in the Department of Computer Science and Artificial Intelligence at the University of the Basque Country, Donostia – San Sebastián, Spain.

Rafael Ramírez-Melendez is an Associate Professor in the Music Technology Group in the Department of Information and Communication Technologies at Pompeu Fabra University, Barcelona, Spain.

Thomas M. Fiore is an Associate Professor of Mathematics at the University of Michigan–Dearborn, USA.

Machine Learning and Music Generation

Edited by
José M. Iñesta, Darrell Conklin, Rafael Ramírez-Melendez, with Thomas M. Fiore as Editor-in-Chief

Routledge
Taylor & Francis Group
LONDON AND NEW YORK

First published 2018
by Routledge
2 Park Square, Milton Park, Abingdon, Oxon, OX14 4RN, UK

and by Routledge
711 Third Avenue, New York, NY 10017, USA

Routledge is an imprint of the Taylor & Francis Group, an informa business

British Library Cataloguing in Publication Data
A catalogue record for this book is available from the British Library

ISBN 13: 978-0-8153-7720-7

Typeset in Times New Roman
by RefineCatch Limited, Bungay, Suffolk

Publisher's Note
The publisher accepts responsibility for any inconsistencies that may have arisen during the conversion of this book from journal articles to book chapters, namely the possible inclusion of journal terminology.

Contents

Citation Information

The chapters in this book were originally published in the *Journal of Mathematics and Music*, Volume 10, Issue 2 (July 2016). When citing this material, please use the original page numbering for each article, as follows:

Editorial
Machine learning and music generation
José M. Iñesta, Darrell Conklin, and Rafael Ramírez-Melendez
Journal of Mathematics and Music, Volume 10, Issue 2 (July 2016), pp. 87–91

Chapter 1
Chord sequence generation with semiotic patterns
Darrell Conklin
Journal of Mathematics and Music, Volume 10, Issue 2 (July 2016), pp. 92–106

Chapter 2
A machine learning approach to ornamentation modeling and synthesis in jazz guitar
Sergio Giraldo and Rafael Ramírez-Melendez
Journal of Mathematics and Music, Volume 10, Issue 2 (July 2016), pp. 107–126

Chapter 3
Analysis of analysis: Using machine learning to evaluate the importance of music parameters for Schenkerian analysis
Phillip B. Kirlin and Jason Yust
Journal of Mathematics and Music, Volume 10, Issue 2 (July 2016), pp. 127–148

Chapter 4
Mapping between dynamic markings and performed loudness: a machine learning approach
Katerina Kosta, Rafael Ramírez-Melendez, Oscar F. Bandtlow, and Elaine Chew
Journal of Mathematics and Music, Volume 10, Issue 2 (July 2016), pp. 149–172

Chapter 5
Data-based melody generation through multi-objective evolutionary computation
Pedro J. Ponce de León, José M. Iñesta, Jorge Calvo-Zaragoza, and David Rizo
Journal of Mathematics and Music, Volume 10, Issue 2 (July 2016), pp. 173–192

Notes on Contributors

Oscar F. Bandtlow is a Senior Lecturer in Applied Mathematics at Queen Mary University of London, UK.

Jorge Calvo-Zaragoza is a Researcher in the Department of Software and Computing Systems at the University of Alicante, Spain.

Elaine Chew is a Professor of Digital Media in the School of Electronic Engineering and Computer Science at Queen Mary University of London, UK.

Darrell Conklin is an Ikerbasque Research Professor in the Department of Computer Science and Artificial Intelligence at the University of the Basque Country, Donostia – San Sebastián, Spain.

Thomas M. Fiore is an Associate Professor of Mathematics at the University of Michigan–Dearborn, USA.

Sergio Giraldo is a Postdoctoral Researcher in the Music Technology Group in the Department of Information and Communication Technologies at Pompeu Fabra University, Barcelona, Spain.

José M. Iñesta is a Professor in the Department of Software and Computing Systems at the University of Alicante, Spain.

Phillip B. Kirlin is an Associate Professor of Computer Science at Rhodes College, Memphis, Tennessee, USA.

Katerina Kosta recently completed a PhD in the School of Electronic Engineering and Computer Science at Queen Mary University of London, UK.

Pedro J. Ponce de León is an Associate Professor in the Department of Software and Computing Systems at the University of Alicante, Spain.

Rafael Ramírez-Melendez is an Associate Professor in the Music Technology Group in the Department of Information and Communication Technologies at Pompeu Fabra University, Barcelona, Spain.

David Rizo is a Lecturer in the Department of Software and Computing Systems at the University of Alicante, Spain.

Jason Yust is an Assistant Professor of Music, Composition, and Music Theory in the College of Fine Arts at Boston University, Massachusetts, USA.

INTRODUCTION

Machine learning and music generation

This special issue on machine learning and music generation features rigorously refereed papers based on selected extended abstracts presented at the 8th International Workshop on Machine Learning and Music (MML2015) held in Vancouver, Canada, in August 2015. The workshop was part of the 21st International Symposium on Electronic Art (ISEA2015).

The annual Machine Learning and Music workshops have established themselves as an active forum bringing together researchers in the area of machine learning applied to music, in order to respond to opportunities and challenges arising from the growth of digital musical data and developments in machine learning and data mining. The MML2015 workshop followed the tradition of previous workshops in Helsinki (MML2008), Bled (MML2009), Florence (MML2010), Athens (MML2011), Edinburgh (MML2012), Prague (MML2013), and Barcelona (MML2014). The 2015 edition was organized in conjunction with the International Symposium on Electronic Art (ISEA), an interdisciplinary event dedicated to creative applications of new technologies in art, interactivity, and electronic and digital media. Therefore MML2015 particularly welcomed contributions on machine learning for music generation and creation.

Computational approaches to music composition and style imitation have engaged musicians, music scholars, and computer scientists since the early days of computing. Music generation research has generally employed one of two strategies: knowledge-based methods that model style through explicitly formalized rules, and data mining methods that apply machine learning to induce statistical models of musical style.

The five papers in this special issue illustrate the range of tasks and design choices in current music generation research applying machine learning techniques and highlighting recurring research issues such as training data, music representation, candidate generation, and evaluation. The papers focus on different aspects of modeling and generating music, including melody (Ponce de León et al. 2016), chord sequences (Conklin 2016), ornamentation (Giraldo and Ramírez 2016), and dynamics (Kosta et al. 2016). Models are induced from audio data (Giraldo and Ramírez 2016; Kosta et al. 2016) or symbolic data (Conklin 2016; Kirlin and Yust 2016; Ponce de León et al. 2016). The papers by Giraldo and Ramírez (2016) and Kirlin and Yust (2016) pay particular attention to the relative contribution of different features employed in representing music data. A challenging issue in computational music generation, especially if fully automated, is the determination of concrete musical events which however cohere with certain more abstract or underlying structures, such as metrical regularities (Ponce de León et al. 2016) or both adjacent and distant repetition (Conklin 2016). Machine learning is employed in generating events and event sequences, for example through model-based prediction (Giraldo and Ramírez 2016; Kosta et al. 2016) or sampling from a statistical model (Conklin 2016), and in evaluating candidate event sequences (Ponce de León et al. 2016) or ranking generated sequences (Conklin 2016) based on a statistical model. The contribution by Kirlin and Yust (2016) presents a step towards learning middleground and background regularities, which will further enable future inductive methods to achieve both stylistic and structural coherence of generated music pieces based on models learned from data.

We guest editors welcome you to this special issue and hope that you find the papers interesting and relevant to your own research.

"Chord Sequence Generation with Semiotic Patterns" by Darrell Conklin

Music derives meaning through repetition. Modeling musical coherence created by internal repetition and reference to earlier material in a piece is a persistent issue in computational music generation, and presents a challenge especially for machine learning methods.

This paper studies repetition in music, and develops a way to represent and generate repetition using statistical models. It develops a pattern representation that explicitly describes relations between events using variables that are instantiated during generation. A *semiotic pattern* describes a formal language of sequences, and sequences satisfying the pattern are generated by sampling from a statistical model. The statistical model is learned on a corpus of chord sequences and is an instance of the *viewpoint* modeling approach used previously with success for prediction, classification, and music generation.

There are two ways to see this work: as a semiotic pattern representation with pattern instances ranked by a statistical model, or as a statistical modeling approach with patterns used to constrain the space of permissible sequences. Whatever the view, the semiotic patterns, which express long-range dependencies, are essential to achieving coherence in generated sequences.

The method is illustrated by generating chord sequences for trance songs. The results show that semiotic patterns and machine learning of statistical models can work together to support music generation which is both internally and stylistically coherent.

"A Machine Learning Approach to Ornamentation Modeling and Synthesis in Jazz Guitar" by Sergio Giraldo and Rafael Ramírez

This paper presents a method to automatically generate expressive jazz guitar performances. A machine learning approach is used to learn when a note in a score is ornamented by a guitarist. Ornaments are applied to the notes of a given jazz tune based on the predictions of such an ornamentation classifier. The symbolic performance produced is then rendered using concatenative synthesis.

The paper describes a system that synthesizes a notated jazz melody using ornamentations of notes that are characteristic for a natural musical interpretation. It relies on a classifier that learns, from a dataset of recorded performances, which notes to ornament and which notes not to ornament. The recorded performances are transcribed and aligned to their respective scores, resulting in a set of data instances where each note in the score is either played as a single note, or as multiple notes. The latter case is referred to as ornamentation. Each of the score notes is represented by a vector of features conveying information such as pitch and duration, information about neighboring notes, and about more global context (key, tempo, etc.). A binary classifier predicts from the feature vectors whether a score note should be ornamented or not. In the case of a predicted ornamentation, an ornamentation is instantiated by adopting the ornamentation of the nearest ornamented neighbor of the note to be ornamented, in the feature space defined by the training data. In case no ornamentation is predicted, the same procedure is applied, but selecting the nearest non-ornamented neighbor. The results are concatenated and transformed in pitch and time, in order to match the pitch and duration of the current note. The result is an audio file of the performed melody.

The experimental part of the paper runs two feature selection algorithms and tests several classifiers for predicting ornamentation. The results show that using more than the best three features does not improve classification accuracy. The authors also conclude that overfitting is minimal using a subset of the training data, at least for some classifiers.

"Analysis of Analysis: Using Machine Learning to Evaluate the Importance of Music Parameters for Schenkerian Analysis" by Phillip B. Kirlin and Jason Yust

This paper reports on experiments with learning probabilities for melodic reductions within a system intended to model Schenkerian analysis, directed at discovering the importance of different features for learning probabilities leading to analyses that most closely match the ground truth.

Several methods to approach Schenkerian analysis from a computer can be found in the literature. In general, they try to systematize expert practice in order to make it computable. Those works are usually outlined as a proof of concept and accordingly evaluated with small corpora or even individual short pieces. One of the most difficult impediments to progress in research on Schenkerian analysis from a computational point of view is the lack of available curated corpora. In previous work, Kirlin introduced a new corpus based on Yust's maximal outerplanar graphs representation that begins to fill this gap.

Using that corpus, the authors have recently introduced an algorithmic technique for analyzing a musical work using a Schenkerian approach. Musical pieces encoded in a symbolic format are described in terms of melodic, harmonic, and metrical features. These 18 features model note triples, and cover aspects such as the scale degree of notes, the harmony present at the onset of the center note, the interval from the left to the right note, and the metrical strength of the left note compared to the center note.

The main contribution of the article presented in this special issue is the use of a machine learning approach to study objectively the influence of the melodic, harmonic, and metrical features for solving a Schenkerian analysis task. Random forests and probabilistic context-free grammars are used in a parsing algorithm that determines the most probable analyses of works in the corpus, which are compared to the manually analyzed works in the corpus.

The result of the study is an ordering of the features by what the authors name *expendability*: how useful is each feature in the context of others. From the results the authors conclude that though all features are significant and necessary, melodic intervals and melodic orientation to a key are the most essential factors for performing a Schenkerian analysis.

"Mapping between Dynamic Markings and Performed Loudness: A Machine Learning Approach" by Katerina Kosta, Rafael Ramírez, Oscar F. Bandtlow, and Elaine Chew

Musicians manipulate sound properties such as pitch, timing, loudness, and timbre in an attempt to communicate emotions, clarify musical structure, or shape the music according to their own intentions. Among these properties, loudness manipulation is one of the main actions for expressivity in music performance. In the context of classical music, loudness specifications are frequently notated in music scores as dynamic markings. However, the meaning of such markings is in general ambiguous and depends on the context in which the markings appear.

This paper investigates the relationship between dynamic markings in music scores and performed loudness by applying machine-learning techniques to induce predictive models of loudness levels corresponding to dynamic markings, and to classify dynamic markings given loudness values. The predictive models are obtained by mining a dataset consisting of 44 recordings of performances of Chopin's Mazurkas each by eight pianists. The results reported in the paper show that loudness values and markings can be predicted when trained across recordings of the same piece, but this is not the case when trained across the pianist's recordings of other pieces. This demonstrates that score features seem to be more important than individual style when modeling loudness choices. The results also show that for some Mazurkas in the dataset loudness values can be predicted more easily than for others. This result is related to the range of the different markings that appear in a piece as well as their position and function in the score in relation to structurally important elements.

"Data-Based Melody Generation through Multi-Objective Evolutionary Computation" by Pedro J. Ponce de León, José M. Iñesta, Jorge Calvo-Zaragoza, and David Rizo

This paper studies the role of statistical and structural descriptors in a genetic system for melody composition. Genetic algorithms are an optimization technique that mimics the biological evolution of living beings and natural selection. A population of individuals, representing different possible solutions, are subjected to crossovers and mutations, and a selection stage decides which individuals might best solve the problem. Those individuals are allowed to procreate a new generation of, supposedly, better individuals until convergence. There are two main difficulties here: the choice of a proper representation of the problem as a "chromosome," and the implementation of the selection process.

In this work, genetic operators are adapted to the chromosomic representation of melodies as *melodic trees*, based on the rhythm structure as a tree. Each bar is represented as a subtree, depending on its meter. The level of a node determines its duration: the root represents the duration of the whole melody, the nodes of the next level represent a division of the upper node that, together, sum up to its duration. Pitch codes are found in the leaves of the tree and any kind of absolute or relative pitch can be used.

The other issue to solve is how to compute the "fitness" of each individual: how every solution is able to actually *solve* the problem at hand, in this case to have an acceptable music composition. The paper uses a combination of different fitness functions by means of a multi-objective optimization method. This is able to combine multiple and diverse values in order to rank the individual in a single fitness space, based on statistical and structural descriptors of the melodies represented by the trees. A number of machine learning techniques are used to rate evolutionary generated music according to the parameters tuned from melodies that supposedly share some common properties such as genre, style, and mood. The authors present a graphical interface prototype that allows one to explore the creative capabilities of the proposed system.

Acknowledgments and Funding

We point out that this special issue would not have been possible without the help and dedication of many experts in the areas of music, mathematics, and machine learning, who provided detailed anonymous reviews of the manuscripts submitted to this special issue. We thank the anonymous referees for their rigorous work, and we thank all authors for their very professional work during the intense writing and revision processes. Special thanks are given to Thomas Fiore who enthusiastically supported this special issue since its inception and provided detailed editorial comments on all of the manuscripts. We would also like to thank Moreno Andreatta for supporting the special issue when it was first proposed, and Kerstin Neubarth for assistance with this editorial article.

We guest editors acknowledge the financial support of the following funding sources: the Ministerio de Economía y Competitividad project TIMuL [No. TIN2013–48152–C2–1–R]; [No. TIN2013–48152–C2–2–R, supported by UE FEDER funds]; the project Lrn2Cre8, which is funded by the Future and Emerging Technologies (FET) programme within the Seventh Framework Programme for Research of the European Commission [FET grant number 610859]; the European Union Horizon 2020 research and innovation programme [grant agreement number 688269].

José M. Iñesta
Darrell Conklin
and
Rafael Ramírez-Melendez

Disclosure statement

No potential conflict of interest was reported by the guest editors.

References

Conklin, Darrell. 2016. "Chord Sequence Generation with Semiotic Patterns." *Journal of Mathematics and Music* 10 (2): 92–106. doi:10.1080/17459737.2016.118172

Giraldo, Sergio, and Rafael Ramírez. 2016. "A Machine Learning Approach to Ornamentation Modeling and Synthesis in Jazz Guitar." *Journal of Mathematics and Music* 10 (2): 107–126. doi:10.1080/17459737.2016.1207814

Kirlin, Phillip B., and Jason Yust. 2016. "Analysis of Analysis: Using Machine Learning to Evaluate the Importance of Music Parameters for Schenkerian Analysis." *Journal of Mathematics and Music* 10 (2): 127–148. doi:10.1080/17459737.2016.1209588

Kosta, Katerina, Rafael Ramírez, Oscar F. Bandtlow, and Elaine Chew. 2016. "Mapping between Dynamic Markings and Performed Loudness: A Machine Learning Approach." *Journal of Mathematics and Music* 10 (2): 149–172. doi:10.1080/17459737.2016.1193237

Ponce de León, Pedro J., José M. Iñesta, Jorge Calvo-Zaragoza, and David Rizo. 2016. "Data-Based Melody Generation through Multi-Objective Evolutionary Computation." *Journal of Mathematics and Music* 10 (2): 173–192. doi:10.1080/17459737.2016.118171

Chord sequence generation with semiotic patterns

Darrell Conklin

Chord sequences play a fundamental role in music analysis and generation. This paper considers the task of chord sequence generation for electronic dance music, specifically for pieces in the sub-genre of trance. These pieces contain repetitive harmonic sequences with a specific semiotic structure and coherence. The paper presents a formal representation for the semiotic structure of chord sequences, and describes a method for ranking the instances of a semiotic pattern using a statistical model trained on a corpus. Examples of generated chord sequences for a full trance piece are presented.

1. Introduction

The visionary enthusiasm of Ada Lovelace (1815–1852) that a mechanical device "might compose elaborate and scientific pieces of music of any degree of complexity or extent" is as motivating and fascinating today as it was pure science fiction in the nineteenth century. From the 1950s, with the first experiments in computational music generation (Brooks et al. 1957; Hiller 1970; Hiller and Isaacson 1959), through to today, the dream and excitement of computational creativity has never faded for computer scientists and musicologists.

Music generation methods can be broadly divided into two categories: *rule-based* methods use specified rules for style emulation and algorithmic composition; *machine learning* methods build generative statistical models from training corpora. A classic problem faced by all methods for music generation is how to appropriately balance general *extraopus* stylistic features with *intraopus coherence* created through reference to earlier music material in a piece. This duality has been referred to variously as *long-term versus short-term models* (Conklin and Witten 1995), *prior knowledge versus on-the-fly knowledge* (Cunha and Ramalho 1999), *extraopus versus intraopus style* (Narmour 1990), *schematic versus contextual probabilities* (Temperley 2014), and *schematic versus veridical expectancies* (Bharucha and Todd 1989).

The coherence problem poses big challenges for machine learning methods, because succinctly modeling intraopus reference is beyond finite-state, and even context-free, grammars.

Figure 1. An instance of the copy language in music. There are two four-event sequences, separated by an arbitrary sequence (indicated by "..."). The events could be any type of music object: notes, motifs, phrases, chords, entire sections, and so on. The network shows that there will exist arbitrarily long-range crossing dependencies that are both identical (solid arrows) and not literal (the dashed arrow).

This can be shown formally by considering the *copy language* (sequences of the form $\boldsymbol{e} \ldots \boldsymbol{e}$ for any sequence $\boldsymbol{e}$) which, despite its simplicity, cannot be generated by even a context-free grammar as it contains an arbitrary number of *crossing dependencies* (see Figure 1). The presence of crossing dependencies that the entire class of *context* models (e.g. Markov models, n-gram models, multiple viewpoint models, hidden Markov models) – which have great practical and computational advantages for machine learning and statistical modeling – cannot handle the coherence problem. This is especially a problem for genres such as electronic dance music which are "unapologetically repetitive" (Garcia 2005).

One way to handle the coherence problem, while retaining the advantages and tractability of context models, is to *adopt the structure* of a template piece, and disallow sequences which do not satisfy that structure (Conklin 2003). To do this, it is necessary to use a pattern formalism that can describe the desired intraopus references in the template piece. Then instances of patterns can be ranked using the statistical model, or in another view, pieces generated by the model can be constrained by the pattern. The pattern captures the intraopus coherence, inherited from the template piece, and the statistical model captures the extraopus style. This idea is developed by Collins et al. (2016), who view a piece as a network of patterns described and related by pitch transposition. In the present paper the idea is generalized and developed further by using *semiotic patterns* representing the references that should be conserved in a generated piece, and by using a *sampling method* to generate large numbers of solutions to the pattern, ranked by their probability according to a context model.

1.1. *Trance and chord sequences*

In music informatics, harmony and chord sequences provide a powerful abstract representation of tonal music, and have a huge importance and impact in music classification (Pérez-Sancho, Rizo, and Iñesta 2009), music analysis (de Clercq and Temperley 2011; Pardo and Birmingham 2002; Rohrmeier 2011), extraction of chord structure from audio (McVicar et al. 2014), and chord sequence prediction (Scholz, Vincent, and Bimbot 2009) and generation (Eigenfeldt and Pasquier 2010). Methods for tonal, polyphonic music generation and melody harmonization are often based on an underlying model of chord transitions (Chuan and Chew 2011; Raczyński, Fukayama, and Vincent 2013; Suzuki and Kitahara 2014). Harmony and chord sequences also play a fundamental role in the understanding of electronic dance music, particularly classically tonal sub-genres of *trance* coming under different labels such as *uplifting trance*, *epic trance*, *progressive trance*, and *euphoric trance*.

Trance songs are characterized by a tempo typically between 135 and 140 bpm in 4/4 metre, have roughly 200 bars of music (approximately 5–7 minutes), and are stitched into longer mixes chosen by a DJ. An uplifting trance song usually contains five distinctly recognizable functional sections, even though it is hard to identify a prototypical form because the ordering and relative length of these sections can vary greatly. The *intro* introduces a tempo, beat, and tonal center, and instruments are gradually layered. A chord sequence loop of either 8 or 16 bars, usually at a harmonic rhythm of two bars per chord, is typically introduced in the *break*, which has an ethereal texture, often with orchestral instruments and without percussion. The break can also introduce a melody, and is where the singing may start in the case of vocal trance. Following the retraction of the texture into near silence, the *build* continues the same loop as the break but again with added layering of instrumentation and intensity. This leads to a typical snare roll and to the climactic and high energy *anthem*, which generally has the same chord loop as the break. A song finishes with an *outro* which retracts the texture in an almost palindromic way to the build of the texture in the intro.

Relying mainly on simple major and minor triadic forms, the chord sequence in these loops is one of the essential components of a trance song. Chord repetition thus occurs on several levels: locally between iterations of the same loop, forming a hypermetrical structure of, for instance, 8, 16, and 32 bar units; distantly (e.g. the outro referencing the loop of the intro); and internally within a loop (still non-adjacent, but at a shorter timescale). A pivotal issue for trance generation is therefore the generation of idiomatic chord loops within globally coherent chord sequences for entire pieces.

This paper develops a representation and generation method that is able to generate chord sequences for trance, though the method has no inherent restriction to this genre. A central feature of the method is the use of *semiotic patterns* (sequences of variables) within the statistical generation process. These semiotic patterns are used to maintain intraopus coherence, as mentioned above a difficult problem in music generation from statistical models and a crucial issue in genres such as electronic dance music that have extensive intraopus repetition of music material.

1.2. *Semiotic analysis*

To represent the structure of a template piece, a new method of *semiotic patterns* is developed, drawing on the field of *semiotic analysis* (Anagnostopoulou 1997; Cambouropoulos and Widmer 2000; Conklin 2006; Cook 1987; Nattiez 1975; Ruwet 1966). This type of music analysis divides a piece into similar segments, called *paradigmatic units*, and expresses the sequence of these units through time. Semiotic analysis is usually applied at the larger motif or phrase level (Bimbot et al. 2012; Cook 1987), but can also be productively applied at small units: the unit of the chord span (for trance anthems, a harmonic rhythm of typically one or two bars). For example, Figure 2 (left) shows an analysis of a short eight-chord sequence, containing five distinct chords (paradigmatic units). Numbers refer to paradigmatic unit (chord) number, and the sequence is read from the analysis like a book: left-to-right, top-to-bottom.

A *pattern* in music informatics, especially in research on pattern discovery from single pieces or from corpora (Conklin 2010a, 2010b; Conklin and Anagnostopoulou 2001; Lartillot 2005; Meredith, Lemström, and Wiggins 2002), typically specifies *what* is conserved among pattern instances. This can be contrasted with a pattern containing *variables* that can be instantiated by any elements of the vocabulary (Angluin 1980). Thus a pattern with variables can specify intraopus references as equality relations between positions arbitrarily distant in a piece. Figure 2 (right) shows such an abstraction, where the concrete paradigmatic units (chords) occurring in a semiotic analysis have been replaced by variables (indicated by underlined letters).

1	2	3	4	5
cm	fm	A♭M		
		A♭M	E♭M	
	fm	A♭M		B♭M

1	2	3	4	5
A	B	C		
		C	D	
	B	C		E

Figure 2. Left: a semiotic analysis of an eight-chord sequence cm, fm, A♭M, A♭M, E♭M, fm, A♭M, B♭M, made up of five distinct chords. Right: its transformation to an analysis where the concrete chords have been replaced by variables, leading to the semiotic pattern ABCCDBCE.

In some earlier music informatics research, patterns with variables have been used (Bel and Kippen 1992; Orlarey et al. 1994): the present paper carries the idea further by allowing patterns to specify equality relations between abstract features, not only concrete events, and Section 3.1 shows how instances can be ranked using a generative statistical model.

2. Representation

In this section the method for representing semiotic patterns in music is developed. After introducing the representation of a music surface in terms of events and sequences, the *viewpoints method* is presented: this method can specify abstractions of events in sequences, and is also used for statistical modeling with these abstractions.

Following the introduction of the viewpoints method, a representation for semiotic patterns containing variables will be developed, and the formal semantics of a pattern will be given by defining its formal language, that is, the set of all sequences instantiating the pattern. Table 1 provides a glossary of notation that will be used throughout this section, and Appendix A provides some further remarks and elaborations on examples of semiotic patterns.

Table 1. Glossary of notation and terminology used in this paper.

Notation	Meaning
ξ	event space
ξ^*	sequence space, free monoid on ξ
$\tau : \xi \to [\tau]$	a viewpoint τ returning values in its codomain $[\tau]$
$\tau(e \mid \cdot)$	viewpoint value of event e in the context of preceding events
$\tau_1 \otimes \tau_2$	linking two viewpoints
$\tau : v$	a feature of an event
δ	diatonic interval between two tonal pitch classes
$\boldsymbol{e}_\ell$	abbreviation for a sequence $e_1 \ldots e_\ell$
$\boldsymbol{e}^k$	k repetitions of a sequence $\boldsymbol{e}$
Φ	a semiotic pattern
μ	a substitution: a function mapping variables to values
$[\![\Phi]\!]_\mu$	extension of pattern Φ using substitution μ
$L(\Phi)$	language described by a pattern Φ: union of all possible extensions
X_i	random variable modeling sequence position i
$\mathbb{P}$	probability
$\mathbb{I}$	information content of a sequence: $\mathbb{I}(\boldsymbol{e}) = -\log_2 \mathbb{P}(\boldsymbol{e})$

2.1. *Events, sequences, and viewpoints*

To represent properties of chords in sequences, we use the method of *viewpoints* (Conklin 2006, 2010a, 2013; Conklin and Witten 1995), which is reviewed here. A *music object* is a primitive type such as a note, a chord, or a complex type such as a sequence or simultaneity of music objects. All music objects have a duration. Within a sequence, an object has an inter-onset interval (time from the onset of the previous event onset time) and is called an *event*. The set ξ, called the *event space*, will be used to denote all events that can be formed from objects of a particular type. The event space is usually, but need not be, taken from events encountered in a training corpus.

An *event sequence* is a sequence formed by concatenating events of the event space ξ, i.e. an event sequence is an element of the free monoid ξ^* on the set ξ, with the empty sequence denoted by Λ. For notational convenience, a sequence $e_1 \ldots e_\ell$ will sometimes be abbreviated as $\boldsymbol{e}_\ell$ (with the e in bold). The notation $\boldsymbol{e}^k$ denotes the event sequence $\boldsymbol{e}$ repeated k times.

A *viewpoint* is a function mapping events to derived features according to the schema

$$\tau : \xi \rightarrow [\tau],$$

where the notation $[\tau]$ is used to refer to the *codomain* of the function τ (the set of possible values that can be returned by the function). The function is partial, therefore it may be undefined ($\perp$) for some events. A viewpoint can refer to a preceding event, or indeed the entire contextual sequence preceding an event: for notational convenience here when an event e_j is notated as an argument to a viewpoint τ, as $\tau(e_j)$, the context $\boldsymbol{e}_{j-1}$ is implied. If necessary to explicitly mention the contextual sequence of an event e, the notation $\tau(e \mid \cdot)$ will be employed, to further emphasize that a viewpoint *assigns a value* based on context to the event.

New viewpoints can be derived from others in various ways: by defining a new function that references existing viewpoints; or more conveniently using higher-order functions called *constructors*. For example, the $\otimes$ constructor links together two viewpoints into a new one which returns pairs of values, one from each component viewpoint:

$$[\tau_1 \otimes \tau_2] = [\tau_1] \times [\tau_2]$$
$$(\tau_1 \otimes \tau_2)(e) = \langle \tau_1(e), \tau_2(e) \rangle,$$

similar to a *direct product* of two generalized interval systems (Lewin 1987), though here the values in the codomain of a viewpoint are not required to be intervals forming a mathematical group structure.

Table 2 presents a small set of viewpoints that can be used to describe chord sequences. In this specification, only major and minor triads are considered (i.e. enhanced chords in the corpus are truncated to a triadic form) and chord roots are represented using tonal (rather than atonal) pitch classes. The event space ξ therefore contains 42 chords: seven pitch classes, each with a flat, natural, or sharp; and the two qualities of major and minor. The crm viewpoint is defined in terms of the diatonic interval between chord roots, using the function δ that returns diatonic interval qualities of major (M), minor (m), augmented (A), diminished (d), and perfect (P), together with seven scale steps. For example: $\delta(\mathsf{C}, \mathsf{F}\sharp) = \mathsf{A4}$, $\delta(\mathsf{C}, \mathsf{D}\flat) = \mathsf{d2}$, and for any pitch class a, $\delta(a, a) = \mathsf{P1}$. A viewpoint crm $\otimes$ cqm linking diatonic root interval and triad quality movement (see Table 2) will be used later in this paper to represent chords for statistical modeling. This linked viewpoint recalls the idea of *uniform triadic transformations* (Hook 2002) though here applied in a diatonic sense. For training statistical models using machine learning, this viewpoint has the desirable property of transposition invariance (see Theorem A.2).

The application of k viewpoints $\tau_1, \ldots, \tau_k$ to an event sequence $e_1 \ldots e_\ell$ may be represented as a $k \times \ell$ solution array where location (i,j) holds the value $\tau_i(e_j)$. Table 3 shows a solution array for a short sequence of eight chords, using the viewpoints presented in Table 2. In this example, time is measured in number of bars, and each chord lasts two bars.

Table 2. A viewpoint specification for chord sequence modeling. The notation [·] is used to indicate the codomain of a viewpoint.

Viewpoint	Description	Codomain
chord	chord	[root] × [qual]
root	chord root	{C,D,E,F,G,A,B} × {♮, ♯, ♭}
qual	major or minor triad quality	{M, m}
ioi	inter onset interval	$\mathbb{N}^+$
dur	duration	$\mathbb{N}^+$
crm	chord diatonic root movement	{d,P,m,M,A} × {1,. . . , 7}
cqm	chord quality movement	{MM,mm,mM,Mm}
crm ⊗ cqm	linked root and quality movement	[crm] × [cqm]

Table 3. A viewpoint solution array for an eight-chord sequence using the viewpoints of Table 2.

chord	AM	f♯m	DM	EM	f♯m	AM	DM	EM
root	A	F♯	D	E	F♯	A	D	E
qual	M	m	M	M	m	M	M	M
ioi	⊥	2	2	2	2	2	2	2
dur	2	2	2	2	2	2	2	2
crm	⊥	M6	m6	M2	M2	m3	P4	M2
cqm	⊥	Mm	mM	MM	Mm	mM	MM	MM
crm ⊗ cqm	⊥	⟨M6,Mm⟩	⟨m6,mM⟩	⟨M2,MM⟩	⟨M2,Mm⟩	⟨m3,mM⟩	⟨P4,MM⟩	⟨M2,MM⟩

2.2. *Semiotic patterns*

In this paper we propose the term *semiotic patterns* to describe music patterns with variables. Semiotic patterns are used to describe sets of music sequences. The components of patterns are sets of features, each feature in turn being a viewpoint name associated with a value or a variable. This section provides a syntax and formal semantics for patterns, then several examples of concrete patterns with their formal interpretations.

2.2.1. *Syntax*

A *feature* of an event is a pair $\tau : v$ comprised of a viewpoint name τ paired with either a *value* or a *variable*: for example, crm:M6 or root:$\underline{\text{A}}$. Note that to distinguish variables from values, variables will be underlined. A *feature set* is a set of features, and a *semiotic pattern* is a sequence of feature sets. The special variable $\underline{\text{X}}$ may be used in a pattern: each occurrence of $\underline{\text{X}}$ is replaced by a fresh variable not occurring elsewhere in the pattern. A *substitution* μ is a function from variables occurring in a pattern to elements of the codomains of viewpoints appearing within the pattern. For convenience, substitutions *preserve values*, meaning that they map all values onto themselves. A substitution μ is *injective* if $\mu(v_1) \neq \mu(v_2)$ for any two distinct variables v_1 and v_2: this is the implicit assumption in semiotic analysis, but is not enforced here by semiotic patterns.

2.2.2. *Formal semantics*

The precise interpretation of a semiotic pattern is presented using an extensional approach, that is, by mapping patterns to sets of concrete sequences in ξ^*. Since patterns may have variables, every different variable substitution will produce a different extension. Thus the extension of a pattern is given by the function $[\![\cdot]\!]_\mu$, which uses specifically the substitution μ.

Each semantic rule in Table 4 handles some syntactic case. First, the empty pattern has a null extension. The second and third semantic rules are recursive definitions: the extension of a pattern Φ concatenated with a feature set (empty in the second rule, or non-empty in the third rule) is all sequences in the extension of the pattern, concatenated with the events that satisfy the feature set. The fourth rule handles the case of bounded repetition of a pattern (which itself might be a repeated pattern). The semantics culminates in the last line of Table 4, which defines ultimately the *language* described by a pattern: the union of all of its extensions, under any possible substitution. If $\boldsymbol{e} \in L(\Phi)$, we say that $\boldsymbol{e}$ is an *instance* of the pattern Φ.

2.2.3. *Examples of semiotic patterns*

The rest of this section presents some examples of semiotic patterns defined using the viewpoint specification of Table 2. The language described by each pattern will be given, and in

Table 4. Syntax and semantics of semiotic patterns.

Syntax	Description and extension
Λ	Empty pattern: $[\![\Lambda]\!]_\mu = \emptyset$
$\Phi\,\{\}$	Pattern Φ (of length $\ell - 1$) followed by an empty feature set: $[\![\Phi\,\{\}]\!]_\mu = \{e_\ell \mid e_{\ell-1} \in [\![\Phi]\!]_\mu \wedge e_\ell \in \xi\}$
$\Phi\,\{f_1,\ldots,f_n,\tau:v\}$	Pattern Φ (of length $\ell - 1$) followed by a non-empty feature set: $[\![\Phi\,\{f_1,\ldots,f_n,\tau:v\}]\!]_\mu = \{e_\ell \mid e_\ell \in [\![\Phi\,\{f_1,\ldots,f_n\}]\!]_\mu \wedge \tau(e_\ell \mid e_{\ell-1}) = \mu(v)\}$
$(\Phi)\{k\}$	Bounded repetition of a pattern Φ: $[\![(\Phi)\{k\}]\!]_\mu = \{e^k \mid e \in [\![\Phi]\!]_\mu\}$
$L(\Phi)$	Language described by a pattern Φ: $L(\Phi) = \bigcup_\mu [\![\Phi]\!]_\mu$

Appendix A some patterns are carried through more thoroughly using the formal semantics of Table 4.

Example 2.1 A pattern describing the set of all sequences comprising two major chords in succession (see Remark A.3 for formal semantics):

$$\{\mathsf{qual} : \mathsf{M}\}\ \{\mathsf{qual} : \mathsf{M}\}.$$

Example 2.2 A pattern describing the set of all sequences comprising two of the same major chords in succession (note the difference from Example 2.1):

$$(\{\mathsf{qual} : \mathsf{M}\})\{2\}.$$

Example 2.3 A pattern describing the set of all sequences with three chords, the first and last being the same, and the second having root F♯:

$$\{\mathsf{chord} : \underline{\mathsf{A}}\}\ \{\mathsf{root} : \mathsf{F}\sharp\}\ \{\mathsf{chord} : \underline{\mathsf{A}}\}.$$

The formal semantics of the above pattern is carried through in Appendix A (Remark A.4).

Example 2.4 The pattern below describes the language of sequences having four chords, repeated three times. The core sequence has some initial chord (variable A), followed by a chord with root C♯ (with B assigned to the variable quality of this chord), followed by the first chord again (which now is seen to be constrained to a minor chord) followed by a chord with the same quality as the second chord:

$$\left(\{\mathsf{chord} : \underline{\mathsf{A}}\} \begin{Bmatrix} \mathsf{root} : \mathsf{C}\sharp \\ \mathsf{qual} : \underline{\mathsf{B}} \end{Bmatrix} \begin{Bmatrix} \mathsf{chord} : \underline{\mathsf{A}} \\ \mathsf{qual} : \mathsf{m} \end{Bmatrix} \{\mathsf{qual} : \underline{\mathsf{B}}\}\right)\{3\}.$$

Example 2.5 It is possible for a pattern Φ to be *unsatisfiable*: with $L(\Phi) = \emptyset$. For example, the language described by the pattern

$$\{\mathsf{chord} : \mathsf{cm}\} \begin{Bmatrix} \mathsf{chord} : \mathsf{EM} \\ \mathsf{crm} : \mathsf{P5} \end{Bmatrix}$$

is empty because it is impossible for an E major chord to have its root a perfect fifth above the root of the chord C minor (see Remark A.5 for formal semantics).

Example 2.6 As a second example of an unsatisfiable pattern, the language described by the pattern

$$\begin{Bmatrix} \mathsf{chord} : \underline{\mathsf{A}} \\ \mathsf{qual} : \mathsf{m} \end{Bmatrix} \begin{Bmatrix} \mathsf{chord} : \underline{\mathsf{A}} \\ \mathsf{qual} : \mathsf{M} \end{Bmatrix}$$

is empty because the pattern states that the same chord (variable $\underline{\mathsf{A}}$) must be present in the first and second positions, but it also states that the first is a minor and the second is a major chord. Therefore there is no substitution to the variable $\underline{\mathsf{A}}$ that satisfies those two conditions.

Henceforth the paper will be focused on patterns that contain only the chord viewpoint, where every feature in the pattern has a variable. For notational convenience such patterns will be abbreviated simply as strings of single-character variables, e.g. the pattern

$$\{\mathsf{chord} : \underline{\mathsf{A}}\} \ \{\mathsf{chord} : \underline{\mathsf{B}}\} \ \{\mathsf{chord} : \underline{\mathsf{A}}\}$$

will be abbreviated to $\underline{\mathsf{ABA}}$.

2.2.4. *Full trance songs*

The first example of a full piece is from a commercially available trance genre template for a digital audio workstation, and the subsequent two are from classic vocal trance works.

Example 2.7 *Uplifting Trance Logic Pro X Template by CJ Stone*, Daw Templates, Germany:

$$(\underline{\mathsf{ABCD}})\{4\} \ (\underline{\mathsf{ABCDABCE}})\{3\} \ (\underline{\mathsf{ABCD}})\{4\} \ (\underline{\mathsf{ABCDABCE}})\{3\}.$$

This piece has a harmonic rhythm of two bars per chord. The sequence has been segmented into four parts, corresponding to the following functional trance sections: *intro*, *build + anthem*, *break*, and finally *build + anthem*. Noting that the pattern has a higher-level structure, it could also be expressed as a nested pattern:

$$\Big((\underline{\mathsf{ABCD}})\{4\} \ \underline{\mathsf{ABCDABCE}} \ (\underline{\mathsf{ABCDABCE}})\{2\} \Big)\{2\}$$

with the three nested segments describing the *intro/break*, *build*, and *anthem*, with the entire structure repeated.

Example 2.8 *Ramelia (Tribute To Amelia). Ram & Susana. Released: Dec 20, 2013, FSOE under exclusive license to Armada Music B.V.*:

$$(\underline{\mathsf{AABA}})\{2\} \ (\underline{\mathsf{ACBA}})\{8\} \ (\underline{\mathsf{CDBA}})\{10\}.$$

This piece has a harmonic rhythm of two bars per chord, and has been segmented into three repeated patterns describing the *intro*, *build + break*, then *build + anthem + outro*.

Example 2.9 *Built to last. Re:Locate Vs Robert Nickson And Carol Lee. Released: Dec 01, 2014, RazNitzanMusic (RNM)*:

$$(\underline{\mathsf{A}})\{96\} \ (\underline{\mathsf{ABCCDBCE}})\{2\} \ (\underline{\mathsf{A}})\{16\} \ (\underline{\mathsf{ABCCDBCE}})\{8\} \ (\underline{\mathsf{A}})\{16\}.$$

This piece has a harmonic rhythm of one bar per chord. The segments represent the five units *intro + break*, *build*, *build* (continued), *anthem*, and *outro*.

3. Statistical models

Semiotic patterns can be used for *analytical* tasks (abstraction of a pattern from a particular piece) but in the rest of this paper the focus is on *generation*: given a semiotic pattern, produce instances of the pattern. This requires producing a substitution, and an event sequence instantiating the pattern under this substitution. Formulated this way, the problem is overly general and there should be a way for preference ranking of the instances of patterns. One way to do this is with a generative statistical model, learned from a corpus, that captures extraopus style and provides probabilities to any sequence of events. Then we can generate sequences from this statistical model while ensuring that they satisfy the semiotic pattern.

3.1. *Viewpoint models*

A *statistical model* for event sequences (of length ℓ) is a multivariate distribution over the sequence $X_1, \ldots, X_\ell$, each X_i being a discrete random variable on the event space ξ. The idea of statistical modeling using *viewpoints* is to generalize the task as one of learning models of *abstract sequences* rather than *concrete events*, thus ameliorating sparse data problems. The theory of viewpoints provides a general statistical modeling method for sequences. This paper will use a specific statistical model using the single chord viewpoint $\mathsf{crm} \otimes \mathsf{cqm}$ (see Table 2). Also for simplicity the presentation will be restricted here to a *first-order* model: one where each event depends on one preceding event.

Consider a viewpoint τ that uses one contextual event, and for two successive events a and b let $v = \tau(b \mid a)$. The conditional probability $\mathbb{P}(b \mid a)$ can be written in the form

$$\begin{aligned}
\mathbb{P}(b \mid a) &= \mathbb{P}(v, b \mid a) \quad \text{since } \mathbb{P}(v \mid a, b) = 1 \\
&= \mathbb{P}(v, b, a) \;/\; \mathbb{P}(a) \quad \text{rule of conditional probability} \\
&= \mathbb{P}(v) \times \mathbb{P}(a \mid v) \times \mathbb{P}(b \mid a, v) \;/\; \mathbb{P}(a) \quad \text{chain rule of probability} \\
&= \mathbb{P}(v) \times \mathbb{P}(b \mid a, v) \quad \text{since } a \text{ and } v \text{ are independent: } \mathbb{P}(a \mid v) = \mathbb{P}(a).
\end{aligned}$$

Now turning specifically to chords and to the viewpoint $\tau = \mathsf{crm} \otimes \mathsf{cqm}$, since b is fully determined by a and v (see Theorem A.2), we have $\mathbb{P}(b \mid a, v) = 1$, and the conditional probability above simplifies to

$$\mathbb{P}(b \mid a) = \mathbb{P}(\tau(b \mid a)). \tag{1}$$

Training a first-order statistical model of the viewpoint $\mathsf{crm} \otimes \mathsf{cqm}$ therefore comprises three steps: transforming all pieces in the training corpus by applying the viewpoint $\mathsf{crm} \otimes \mathsf{cqm}$ to every chord; then, for each $v \in [\mathsf{crm} \otimes \mathsf{cqm}]$ estimating the probability $\mathbb{P}(v)$ as $n(v)/n$, where $n(v)$ is the number of chords in the corpus having the value v, and n is the number of chords in the corpus. Finally, a transition matrix over all concrete events can be compiled according to equation (1).

Given a first-order model as defined above, the probability of a chord sequence $\boldsymbol{e}_\ell$ is

$$\mathbb{P}(e_1 \ldots e_\ell) = \mathbb{P}(e_1) \times \prod_{i=2}^{\ell} \mathbb{P}(e_i \mid e_{i-1}),$$

with the first term $\mathbb{P}(e_1)$ modeled using a uniform distribution over ξ. Sequence probabilities can quickly become very small, so it can be convenient to work with logarithms: the *information content* of a sequence $\boldsymbol{e}$ is defined as $\mathbb{I}(\boldsymbol{e}) = -\log_2 \mathbb{P}(\boldsymbol{e})$.

3.2. *Sampling*

Consider a statistical model $X_1, \ldots, X_\ell$, each X_i a random variable on the event space ξ. In general, the distribution of sequences of length ℓ is unknown, and to enumerate all $|\xi|^\ell$ sequences would be prohibitive in time and space. Therefore *sampling* algorithms must be used to partially recover the sequence probability distribution. Subsequently, sequences can be drawn as desired from chosen regions of the distribution. In general, high probability (low information content) sequences will usually be preferred because they are assumed to contain similar patterns and regularity as the underlying corpus, better conforming to the extraopus style.

The *random walk* method for sequence generation operates by drawing an event e_i from the distribution of the conditional random variable $X_i \mid X_{i-1} = e_{i-1}$ (in the case of a first-order model), concatenating this element to the sequence, and repeating the process until a desired sequence length (or duration) is reached. Single sequences generated using random walk do not optimize the information content (even when following the highest probability transitions at each step: see Remark A.1), and will usually fall around the mean of the sequence distribution. Due to its theoretically complete coverage of the sequence space, the random walk method can be used for sampling, simply by iterating it many times while tabulating the generated sequences and their probabilities (Whorley and Conklin 2015).

The random walk method can also be restricted to the generation of sequences in $L(\Phi)$ for a given pattern Φ. At each step of the walk the possible events are filtered for consistency with the substitution assigned to previous events, and also with the variables at the corresponding position in the semiotic pattern. Only consistent events are considered for continuing the current partial sequence, and a fresh random walk is initiated if no continuation exists. If desired, a partial substitution can be specified, conveniently representing unary constraints (Pachet, Roy, and Barbieri 2011) on the pattern variables.

4. Chord sequence generation for trance

The method described in Section 3.1 was used to train a crm $\otimes$ cqm viewpoint model on a small corpus of 150 anthem loops derived from real trance pieces. Key information, often ambiguous in trance and even more so when restricted just to the anthem section, has not been encoded. The loop lengths average approximately six chords per loop. Chord duration has been encoded, but not used for this study. Most loops have a harmonic rhythm of either one or two bars per chord. During model training, since the sequences represent loops, for every corpus sequence its first chord is concatenated to the sequence before training.

A professional trance template (*Uplifting Trance Logic Pro X Template by Insight*, Daw Templates, Germany) was used to test the generation method. This piece can be described by the semiotic pattern

$$(\underline{\mathsf{ABCD}})\{2\}\ (\underline{\mathsf{AXXX}})\{11\}\ (\underline{\mathsf{ABCD}})\{4\}\ \underline{\mathsf{A}}$$

with the segments representing the units *intro*, *break* + *build*, and *anthem* + *outro*.

For all experiments the variable $\underline{\mathsf{A}}$ was fixed to an A minor chord. The sequence distribution was sampled using 1,000,000 iterations of random walk (Section 3.2), and Figure 3 shows a histogram of information content for approximately 150,000 distinct instances of the pattern. For a given semiotic pattern Φ, the probability mass for sequences in $L(\Phi)$ may not sum to unity due to the constraints imposed by the pattern, therefore in Figure 3 the information content has been normalized. It can be seen that a large portion of the normalized probability mass is within the top ten sequences (left inset of Figure 3), and most sampled sequences have exceedingly low normalized probabilities.

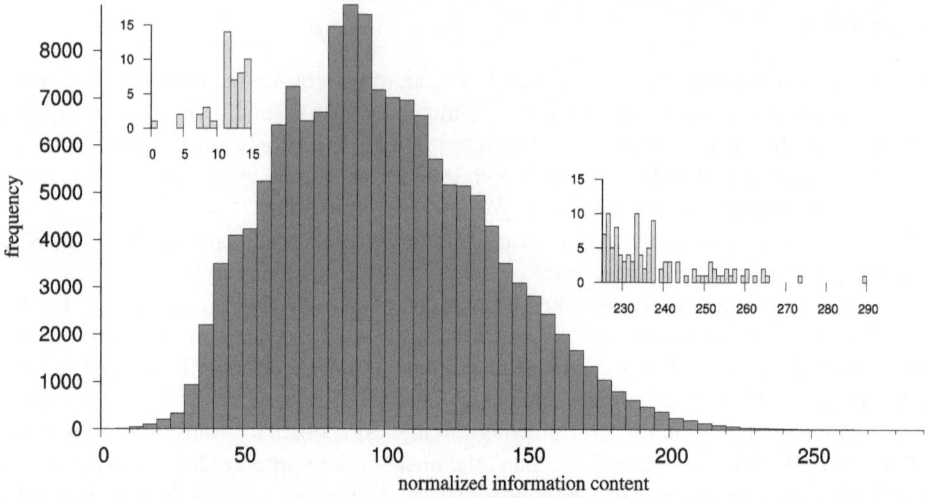

Figure 3. Information content for the semiotic pattern $\underline{(ABCD)}\{2\}$ $\underline{(AXXX)}\{11\}$ $\underline{(ABCD)}\{4\}$ $\underline{A}$, sampled using 1,000,000 iterations of random walk, producing approximately 150,000 distinct instances of the pattern. Information content on the horizontal axis uses bins of width 5, and the vertical axis shows the number of distinct sequences sampled into each bin. The inset histograms focus on the tails of the distribution.

Table 5. Example sequences from high-, mid-, and low-probability solutions, drawn from three regions of the distribution shown in Figure 3. Though full expanded sequences were generated, in the table loops have been compressed. The information content $\mathbb{I}(\cdot)$ is normalized.

$\mathbb{I}(\cdot)$	(A	B	C	D)	(A	X	X	X)	(A	B	C	D)	A
0.197	am	FM	CM	GM	am	FM	CM	GM	am	FM	CM	GM	am
4.323	am	em	CM	GM	am	FM	CM	GM	am	em	CM	GM	am
9.705	am	GM	am	GM	am	FM	CM	GM	am	GM	am	GM	am
119.14	am	dm	em	GM	am	em	DM	AM	am	dm	em	GM	am
119.14	am	CM	dm	gm	am	em	DM	AM	am	CM	dm	gm	am
133.14	am	CM	D♭M	B♭M	am	FM	dm	GM	am	CM	D♭M	B♭M	am
264.02	am	FM	AM	gm	am	BM	GM	AM	am	FM	AM	gm	am
264.18	am	DM	C♯M	AM	am	em	GM	BM	am	DM	C♯M	AM	am
289.04	am	GM	am	BM	am	gm	CM	AM	am	GM	am	BM	am

Table 5 shows some sequences generated in three ranges of the sequence distribution (Figure 3): three sequences within the top ten, three around the mean, and three within the bottom ten. Though no tonality or key information is specified by the model, sequences in the high probability range are tonally consistent, for example the first sequence uses the chords i, VI, III, ♭VII in A minor (vi, IV, I, V in C major), a typical trance chord progression. The second two sequences of Table 5 are also interesting and quite usable for trance. In the middle range of information content the sequences, while having locally acceptable chord transitions, do not follow typical diatonic harmony. Though not as immediately applicable to trance as the top range, they can still be interesting to explore. For example, each of the first two sequences in the middle range has an interesting ⟨P1,Mm⟩ transition that is in fact at the border of the build and anthem. The third contains a ♭II–i progression in A minor, which imparts a Phrygian flavor to the sequence. Finally, sequences in the bottom range of Table 5 comprise mainly poor local transitions and are unusable for trance or indeed any tonal music, even though they are also constrained by the semiotic pattern. This shows the power of using a statistical model to rank instances of a semiotic pattern.

5. Discussion

Semiotic patterns are a powerful representation for the expression of intraopus repetition and reference relations, even those distant in a piece. This paper has developed and applied an approach for generating chord sequences from a statistical model, while maintaining a specified semiotic structure. Since context models cannot alone generate coherent sequences, a semiotic pattern that explicitly represents coherence relations and long-range dependencies helps to guide the generation towards tonally consistent repeated sequences for trance.

Looking back, this work could have been presented in one of two ways: as a statistical modeling approach, with patterns providing constraints on the sequences generated; or as a knowledge representation approach, with statistical models used to rank the instances of patterns. The latter discourse was chosen here to emphasize the more fundamental importance of representing coherence, repetition, and looping in music generation, especially in trance, and the secondary importance of the actual statistical model.

The evaluation of music generated by statistical models is always a fascinating issue and naturally any final determination must be made by human listeners. Nevertheless, this paper has shown that information content can provide a reasonable ranking method for instances of semiotic patterns, in that high probability sequences tend to be tonally consistent and natural for trance.

Though this paper has discussed just a few chord viewpoints, semiotic patterns are very general, and indeed any feature that can be represented using chord viewpoints can be used. There is no inherent restriction to chord sequences, and any type of musical sequence can be modeled using this method. Finally the semiotic patterns, here provided prior to sampling, can be learned from a corpus if available, and even more interestingly generated from a statistical model.

Acknowledgments

Thanks to Kerstin Neubarth for valuable discussions on the research and the manuscript, to Louis Bigo for transcription of trance anthems, and to the reviewers and Thomas Fiore for helpful comments on the manuscript.

Disclosure statement

No potential conflict of interest was reported by the author(s).

Funding

This research is supported by the project Lrn2Cre8, which is funded by the Future and Emerging Technologies (FET) programme within the Seventh Framework Programme for Research of the European Commission [FET grant number 610859].

References

Anagnostopoulou, C. 1997. "Lexical Cohesion in Linguistic and Musical Discourse." In *Proceedings of the 3rd Triennial Conference of the European Society for the Cognitive Sciences of Music (ESCOM)*, Uppsala, Sweden, 655–660.

Angluin, D. 1980. "Finding Patterns Common to a Set of Strings." *Journal of Computer and System Sciences* 21 (1): 46–62.

Bel, B., and J. Kippen. 1992. "Bol Processor Grammars." In *Understanding Music with AI*, edited by M. Balaban, K. Ebcioğlu, and O. E. Laske, 366–400. Menlo Park, CA: AAAI Press.

Bharucha, J. J., and P. M. Todd. 1989. "Modeling the Perception of Tonal Structure with Neural Nets." *Computer Music Journal* 13 (4): 44–53.

Bimbot, F., E. Deruty, G. Sargent, and E. Vincent. 2012. "Semiotic Structure Labeling of Music Pieces: Concepts, Methods and Annotation Conventions." In *Proceedings of the 13th International Society for Music Information Retrieval Conference (ISMIR 2012)*, Porto, Portugal, 235–240.

Brooks, F. P., A. L. Hopkins Jr, P. G. Neumann, and W. V. Wright. 1957. "An Experiment in Musical Composition." *IRE Transactions on Electronic Computers* EC-6 (3): 175–182.

Cambouropoulos, E., and G. Widmer. 2000. "Automated Motivic Analysis via Melodic Clustering." *Journal of New Music Research* 29 (4): 347–370.

Chuan, C.-H., and E. Chew. 2011. "Generating and Evaluating Musical Harmonizations that Emulate Style." *Computer Music Journal* 35 (4): 64–82.

Collins, T., R. Laney, A. Willis, and P. H. Garthwaite. 2016. "Developing and Evaluating Computational Models of Musical Style." *Artificial Intelligence for Engineering Design, Analysis and Manufacturing* 30 (1): 16–43.

Conklin, D. 2003. "Music Generation from Statistical Models." In *Proceedings of the AISB Symposium on Artificial Intelligence and Creativity in the Arts and Sciences*, 7–11 April 2003, Aberystwyth, Wales, 30–35. http://citeseerx.ist.psu.edu/viewdoc/download?doi = 10.1.1.3.2086&rep = rep1&type = pdf.

Conklin, D. 2006. "Melodic Analysis with Segment Classes." *Machine Learning* 65 (2-3): 349–360.

Conklin, D. 2010a. "Discovery of Distinctive Patterns in Music." *Intelligent Data Analysis* 14 (5): 547–554.

Conklin, D. 2010b. "Distinctive Patterns in the First Movement of Brahms' String Quartet in C Minor." *Journal of Mathematics and Music* 4 (2): 85–92.

Conklin, D. 2013. "Multiple Viewpoint Systems for Music Classification." *Journal of New Music Research* 42 (1): 19–26.

Conklin, D., and C. Anagnostopoulou. 2001. "Representation and Discovery of Multiple Viewpoint Patterns." In *Proceedings of the International Computer Music Conference (ICMC 2001)*, Havana, Cuba, 479–485.

Conklin, D., and I. Witten. 1995. "Multiple Viewpoint Systems for Music Prediction." *Journal of New Music Research* 24 (1): 51–73.

Cook, N. 1987. *A Guide to Musical Analysis*. Oxford, UK: Oxford University Press.

Cunha, U. S., and G. Ramalho. 1999. "An Intelligent Hybrid Model for Chord Prediction." *Organised Sound* 4 (2): 115–119.

de Clercq, T., and D. Temperley. 2011. "A Corpus Analysis of Rock Harmony." *Popular Music* 30 (1): 47–70.

Douthett, J., and J. Hook. 2009. "Formal Diatonic Intervallic Notation." In *Mathematics and Computation in Music*, edited by E. Chew, A. Childs, and C.-H. Chuan, 104–114. Berlin: Springer-Verlag.

Eigenfeldt, A., and P. Pasquier. 2010. "Realtime Generation of Harmonic Progressions Using Controlled Markov Selection." In *Proceedings of the 1st International Conference on Computational Creativity (ICCC 2010)*, Lisbon, Portugal, 16–25.

Garcia, L.-M. 2005. "On and On: Repetition as Process and Pleasure in Electronic Dance Music." *Music Theory Online* 11 (4): SEP. http://www.mtosmt.org/issues/mto.05.11.4/mto.05.11.4.garcia_frames.html.

Hiller, L. 1970. "Music Composed with Computers – A Historical Survey." In *The Computer and Music*, edited by H. B. Lincoln, 42–96. Ithaca, NY: Cornell University Press.

Hiller, L., and L. M. Isaacson. 1959. *Experimental Music*. New York: McGraw–Hill.

Hook, J. 2002. "Uniform Triadic Transformations." *Journal of Music Theory* 46 (12): 57–126.

Lartillot, O. 2005. "Efficient Extraction of Closed Motivic Patterns in Multi-Dimensional Symbolic Representations of Music." In *Proceedings of the 6th International Conference on Music Information Retrieval (ISMIR 2005)*, London, 191–198.

Lewin, D. 1987. *Generalized Musical Intervals and Transformations*. New Haven, CT: Yale University Press.

McVicar, M., R. Santos-Rodriguez, Y. Ni, and T. De Bie. 2014. "Automatic Chord Estimation from Audio: A Review of the State of the Art." *IEEE/ACM Transactions on Audio, Speech, and Language Processing* 22 (2): 556–575.

Meredith, D., K. Lemström, and G. Wiggins. 2002. "Algorithms for Discovering Repeated Patterns in Multidimensional Representations of Polyphonic Music." *Journal of New Music Research* 31 (4): 321–345.

Narmour, E. 1990. *The Analysis and Cognition of Basic Melodic Structures: The Implication-Realization Model*. Chicago, IL: University of Chicago Press.

Nattiez, J.-J. 1975. *Fondements d'une Sémiologie de la Musique*. Paris: Union Générale d'Editions.

Orlarey, Y., D. Fober, S. Letz, and M. Bilton. 1994. "Lambda Calculus and Music Calculi." In *Proceedings of the International Computer Music Conference (ICMC 1994)*, Aarhus, Denmark, 253–250.
Pachet, F., P. Roy, and G. Barbieri. 2011. "Finite-Length Markov Processes with Constraints." In *Proceedings of the 22nd International Joint Conference on Artificial Intelligence (IJCAI 2011)*, Barcelona, Spain, 635–642.
Pardo, B., and W. P. Birmingham. 2002. "Algorithms for Chordal Analysis." *Computer Music Journal* 26 (2): 27–49.
Pérez-Sancho, C., D. Rizo, and J.-M. Iñesta. 2009. "Genre Classification Using Chords and Stochastic Language Models." *Connection Science* 20 (23): 145–159.
Raczyński, S. A., S. Fukayama, and E. Vincent. 2013. "Melody Harmonization with Interpolated Probabilistic Models." *Journal of New Music Research* 42 (3): 223–235.
Rohrmeier, M. 2011. "Towards a Generative Syntax of Tonal Harmony." *Journal of Mathematics and Music* 5 (1): 35–53.
Ruwet, N. 1966. "Méthodes d'analyse en musicologie." *Revue Belge de Musicologie* 20 (1/4): 65–90.
Scholz, R., E. Vincent, and F. Bimbot 2009. "Robust Modeling of Musical Chord Sequences Using Probabilistic *n*-Grams." In *IEEE International Conference on Acoustics, Speech and Signal Processing (ICASSP 2009)*, Taipei, Taiwan, 53–56.
Suzuki, S., and T. Kitahara. 2014. "Four-Part Harmonization Using Bayesian Networks: Pros and Cons of Introducing Chord Nodes." *Journal of New Music Research* 43 (3): 331–353.
Temperley, D. 2014. "Information Flow and Repetition in Music." *Journal of Music Theory* 58 (2): 155–178.
Whorley, R. P., and D. Conklin 2015. "Improved Iterative Random Walk for Four-Part Harmonisation." In *Mathematics and Computation in Music, Proceedings of the 5th International Conference (MCM 2015)*, 22–25 June 2015, London, edited by T. Collins, D. Meredith, and A. Volk, Vol. 9110 of the series *Lecture Notes in Computer Science*, 64–70. Cham, Switzerland: Springer International Publishing.

Appendix A. Further remarks on semiotic patterns

Remark A.1 Given a statistical model, the highest probability sequence of length ℓ

$$\arg\max_{e_1 \dots e_\ell} \mathbb{P}(e_1 \dots e_\ell)$$

is not necessarily the sequence $e_1 \dots e_\ell$ produced by selecting the highest probability transition at each step, i.e. where

$$e_i = \arg\max_x \mathbb{P}(X_i = x \mid X_{i-1} = e_{i-1}).$$

For a concrete counterexample, consider the following transition matrix:

$$\begin{array}{c} \\ a \\ b \end{array} \begin{array}{c} \begin{array}{cc} a & b \end{array} \\ \begin{pmatrix} 0.6 & 0.4 \\ 1.0 & 0 \end{pmatrix} \end{array}$$

and suppose that $\mathbb{P}(X_1 = a) = 0.7$. Consider $\ell = 3$. By enumeration of the eight possible sequences of length 3, it can be seen that the highest probability sequence is aba with $\mathbb{P}(aba) = 0.7 \times 0.4 \times 1.0 = 0.280$, but the sequence produced by selecting the highest probability transition at each step is aaa with $\mathbb{P}(aaa) = 0.7 \times 0.6 \times 0.6 = 0.252$.

THEOREM A.2 *The viewpoint* crm ⊗ cqm *is invariant under diatonic transpositions.*

Proof The cqm viewpoint does not depend on the chord root so is naturally transposition invariant. Define $\mathsf{crm}(b \mid a) = \delta(\mathsf{root}(a), \mathsf{root}(b))$. Then $[\mathsf{crm}]$ is a group with addition of diatonic intervals as the group operation and with the identity element P1 (Douthett and Hook [2009] provide a derivation of the group structure). Furthermore for all $a \in \xi$ and $i \in [\mathsf{crm}]$ there is a unique $b \in \xi$ such that $\mathsf{crm}(b \mid a) = i$. Therefore $\langle \xi, [\mathsf{crm}], \mathsf{crm} \rangle$ is a *generalized interval system* (Lewin 1987) and can be associated with a group of transpositions acting on events, by associating with each $i \in [\mathsf{crm}]$ a transposition t_i defined by $\mathsf{crm}(t_i(a) \mid a) = i$. Assuming that $\mathsf{crm}(b \mid a) = i$, it follows that $b = t_i(a)$, and therefore $\mathsf{crm}(b \mid a) = i = \mathsf{crm}(t_i(b) \mid b) = \mathsf{crm}(t_i(b) \mid t_i(a))$. ■

A further noteworthy point of Theorem A.2 is that it illustrates the equivalence between viewpoints and generalized interval systems (Lewin 1987), which holds for any viewpoint having a group structure.

Remark A.3 –Formal semantics of pattern Example 2.1

$$\begin{aligned}\llbracket\{\mathsf{qual}:\mathsf{M}\}\ \{\mathsf{qual}:\mathsf{M}\}\rrbracket_{\mu} &= \{e_2 \mid e_1 \in \llbracket\{\mathsf{qual}:\mathsf{M}\}\rrbracket_{\mu} \wedge \mathsf{qual}(e_2)=\mathsf{M}\}\\ &= \text{all sequences } e_1e_2, \text{where}\\ &\quad \mathsf{qual}(e_1)=\mathsf{M}, \text{and } \mathsf{qual}(e_2)=\mathsf{M}.\end{aligned}$$

The pattern therefore describes a sequence of two major chords.

Remark A.4 –Formal semantics of pattern Example 2.3

$$\begin{aligned}\llbracket\{\mathsf{chord}:\underline{\mathsf{A}}\}\ \{\mathsf{root}:\mathsf{F}\sharp\}\ \{\mathsf{chord}:\underline{\mathsf{A}}\}\rrbracket_{\mu} &= \text{all sequences } e_1e_2e_3, \text{where}\\ &\quad \mathsf{chord}(e_1)=\mu(\underline{\mathsf{A}}), \text{and}\\ &\quad \mathsf{root}(e_2)=\mathsf{F}\sharp, \text{and}\\ &\quad \mathsf{chord}(e_3)=\mu(\underline{\mathsf{A}}).\end{aligned}$$

The pattern describes three events $e_1e_2e_3$ where $\mathsf{chord}(e_1)=\mathsf{chord}(e_3)$ according to the presence of the semiotic variable $\underline{\mathsf{A}}$ in the first and third positions.

Remark A.5 –Formal semantics of pattern Example 2.5

$$\begin{aligned}\left\llbracket\{\mathsf{chord}:\mathsf{cm}\}\left\{\begin{matrix}\mathsf{chord}:\mathsf{EM}\\ \mathsf{crm}:\mathsf{P5}\end{matrix}\right\}\right\rrbracket_{\mu} &= \text{all sequences } e_1e_2, \text{where}\\ &\quad \mathsf{chord}(e_1)=\mathsf{cm}, \text{and } \mathsf{chord}(e_2)=\mathsf{EM},\\ &\quad \text{but } \delta(\mathsf{root}(\mathsf{cm}),\mathsf{root}(\mathsf{EM}))=\mathsf{M3}, \text{therefore}\\ &\quad \mathsf{crm}(e_2 \mid e_1)\neq \mathsf{P5},\end{aligned}$$

therefore the pattern is unsatisfiable.

A machine learning approach to ornamentation modeling and synthesis in jazz guitar

Sergio Giraldo and Rafael Ramírez-Melendez

We present a machine learning approach to automatically generate expressive (ornamented) jazz performances from un-expressive music scores. Features extracted from the scores and the corresponding audio recordings performed by a professional guitarist were used to train computational models for predicting melody ornamentation. As a first step, several machine learning techniques were explored to induce regression models for timing, onset, and dynamics (i.e. note duration and energy) transformations, and an ornamentation model for classifying notes as ornamented or non-ornamented. In a second step, the most suitable ornament for predicted ornamented notes was selected based on note context similarity. Finally, *concatenative synthesis* was used to automatically synthesize expressive performances of new pieces using the induced models. Supplemental online material for this article containing musical examples of the automatically generated ornamented pieces can be accessed at doi:10.1080/17459737.2016.1207814 and https://soundcloud.com/machine-learning-and-jazz. In the Online Supplement we present an example of the musical piece *Yesterdays* by Jerome Kern, which was modeled using our methodology for expressive music performance in jazz guitar.

1. Introduction

Performance actions (PAs) can be defined as musical resources used by musicians to add expression when performing a musical piece, which consist of variations in timing, pitch, and energy. In the same context, *ornamentation* can be considered as an expressive musical resource used to embellish and add expression to a melody. In the past, music expression has been mostly studied in the context of classical music, e.g. Puiggròs et al. (2006), in particular classical piano music, e.g. Widmer and Tobudic (2003). Contrary to classical music scores, performance annotations (e.g. ornaments and *articulations*) are seldom indicated in popular music (e.g. jazz music) scores, and it is up to the performer to include them based on his/her musical background. Therefore, in popular music it may not always be possible to characterize ornaments with the archetypical classical music conventions (e.g. trills and appoggiaturas). Several approaches have been proposed to generate expressive performances in jazz saxophone music, e.g. Arcos, De Mantaras, and Serra (1998), Ramírez and Hazan (2006), and Grachten (2006). Ramírez and Hazan (2006) describe a method to predict ornamentation (among other performance actions). Grachten (2006) detects ornaments of multiple notes to render expressive-aware tempo transformations. Other methods are able to recognize and characterize ornamentation in popular music, e.g. Gómez et al. (2011) and Perez et al. (2008). However, due to the complexity of free ornamentation, most of these approaches study ornamentation in constrained settings, for instance by restricting the study to one-note or notated trills ornamentations, e.g. Puiggròs et al. (2006). Based on our previous studies in expressive performance modeling in jazz music (Giraldo 2012; Giraldo and Ramírez 2014, 2015a, 2015b, 2015c, 2015d), this article presents a system for automatically predicting and synthesizing expressive jazz guitar music performances with unrestricted ornamentation.

The aim of this work is twofold: (1) to train computational models of music expression using recordings of a professional jazz guitar player, and (2) to synthesize expressive ornamented performances from inexpressive scores. The general framework of the system is depicted in Figure 1. In order to train a jazz guitar ornamentation model, we recorded a set of 27 jazz standards performed by a professional jazz guitarist. We extracted symbolic features from the scores using information on each note, information on the neighboring notes, and information related to the musical context. The performed pieces were automatically transcribed by applying note segmentation based on pitch and energy information. After performing score-to-performance alignment, using *dynamic time warping* (DTW), we calculated performance actions by measuring the deviations between performed notes and their respective parent notes in the score. For model evaluation, the data set was split using a leave-one-piece-out approach in which each piece was in turn used as test set, using the remaining pieces as training set. The expressive

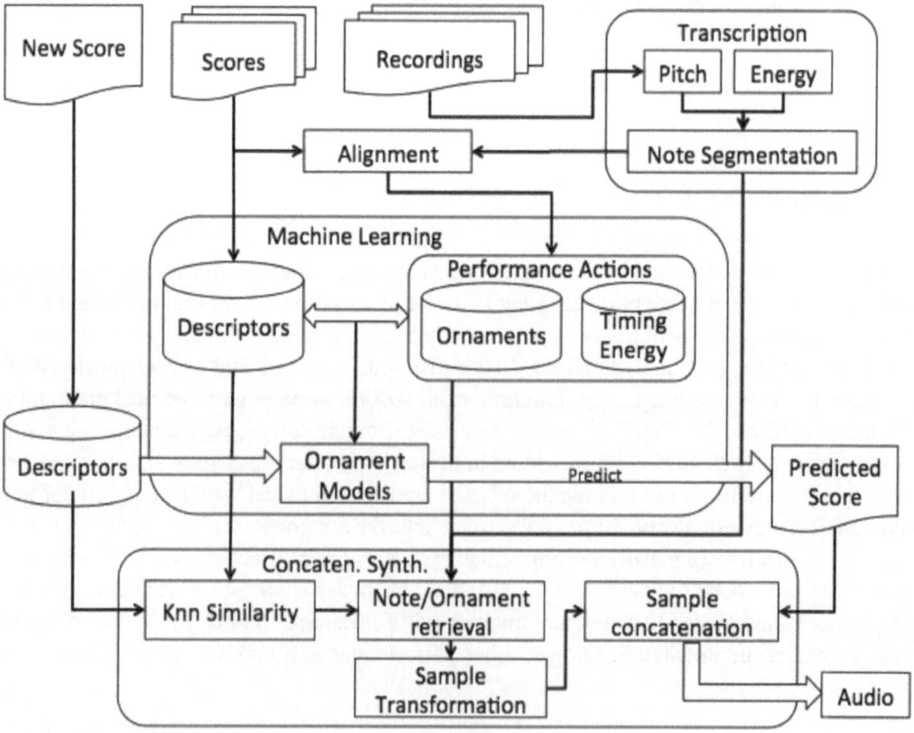

Figure 1. General framework for jazz guitar ornament modeling.

actions considered in this article were duration, onset, energy, and ornamentation transformations. *Concatenative synthesis* was used to synthesize new ornamented jazz melodies using samples of adapted notes/ornaments from the segmented audio recordings.

The rest of the article is organized as follows. Section 2 surveys related work. Section 3 describes data acquisition. Section 4 presents our machine learning approach to ornament prediction. Section 5 describes the audio synthesis method. Section 6 reports on the results, and finally, Section 7 presents some conclusions and future work.

2. Related work

Expressive music performance studies the micro variations a performer introduces (voluntary or involuntary) when performing a musical piece to add expression. Several studies investigating this phenomenon have been conducted, e.g. Gabrielsson (1999, 2003) and Palmer (1997). Computational approaches to studying expressive music performance have been proposed in which data are extracted from real performances and used to formalize expressive models for different aspects of performance – for an overview see Goebl et al. (2008). *Computational systems for expressive music performance* (CEMP) are often targeted at automatically generating human-like performances by introducing variations in timing, energy, and articulation (Kirke and Miranda 2013).

Two main approaches have been used to model expression computationally. On one hand, expert-based systems obtain their rules manually from music experts. A relevant example is the work of the KTH group (Bresin and Friberg 2000; Friberg, Bresin, and Sundberg 2006; Friberg 2006). Their *Director Musices* system incorporates rules for tempo, dynamic, and articulation transformations. Other examples of manually generated expressive systems are the Hierarchical Parabola Model (Todd 1989, 1992, 1995) and the work of Johnson (1991), who developed a rule-based expert system to determine expressive tempo and articulation for Bach's fugues from *The Well-Tempered Clavier*. The rules were obtained from two expert performers. On the other hand, machine-learning-based systems obtain their expressive models from real music performance data by measuring the deviations of a human performance with respect to a neutral or *robotic* performance, using computational learning tools. For example, neural networks were used by Bresin (1998) to model piano performances, and by Camurri, Dillon, and Saron (2000) to model emotional flute performances. Rule-based learning algorithms were used by Widmer (2003) to cluster piano performance rules. Other piano expressive performance systems worth mentioning are the ESP piano system by Grindlay (2005), which utilizes Hidden Markov Models, and the generative performance system of Miranda, Kirke, and Zhang (2010), which uses genetic algorithms to construct tempo and dynamic curves.

Most of the proposed expressive music systems are targeted at classical piano music. More recently, there have been several approaches to computationally modeling expressive performance in popular music by applying machine learning techniques. Arcos, De Mantaras, and Serra (1998) report on *SaxEx*, a performance system capable of generating expressive solo saxophone performances in jazz, based on case-based reasoning. Ramírez and Hazan (2006) compare different machine learning techniques to obtain jazz saxophone performance models capable of both automatically synthesizing expressive performances and explaining expressive transformations. Grachten (2006) applies dynamic programming using an extended version of edit distance and case-based reasoning to detect multiple note ornaments and render expressive-aware tempo transformations for jazz saxophone music. In previous work (Giraldo 2012; Giraldo and Ramírez 2015a, 2015b, 2015c), ornament characterization in jazz guitar performances is accomplished using machine learning techniques to train models for note ornament prediction.

3. Data acquisition

In this study, the data set consisted of 27 jazz standard audio recordings (resulting in a total of 1368 notes) recorded by a professional jazz guitarist, and their corresponding music scores. Each note in the score of the recorded pieces was characterized by a set of 30 descriptors. The music scores and audio recordings were analyzed as explained in Sections 3.1 and 3.2, respectively.

3.1. *Score analysis*

Music scores were obtained from commercially available compilations of jazz scores (The Real Book Series). Selected scores were rewritten using an open source software for music notation, and saved into *MusicXML* format. The MusicXML format allows not only information about the notes (pitch, onset, and duration) to be stored, but also other relevant information for note description such as chords, key, and tempo (among others).

3.1.1. *Feature extraction*

Feature extraction was performed following an approach similar to that of Giraldo (2012), in which each note is characterized by its *nominal*, *neighboring*, and *contextual* properties.

- *Nominal* descriptors refer to the intrinsic properties of score notes (e.g. pitch, duration, and onset). Duration and onsets were described both in beats and seconds, as the duration in seconds depends on the tempo of the piece. For example the choice of ornamenting two different notes from different pieces with quarter note duration (beats) may differ if the pieces are played at slow and fast tempos. The energy descriptor refers to the loudness of the note, which in MIDI format is measured as velocity (how fast a piano key was pressed).
- Given a particular note, its *neighboring* descriptors refer to the properties of its neighboring notes, e.g. previous/next interval, previous/next duration ratio, previous/next inter-onset interval. In this work, only one previous and one following note were considered. Inter-onset distance (Giraldo and Ramírez 2015b) refers to the onset difference between two consecutive notes.
- *Contextual* descriptors refer to the musical *context* in which the note occurs, e.g. tempo, chord, and key. The phrase descriptor (Giraldo and Ramírez 2015b) refers to the note position within a phrase: initial, middle, or end. Phrase descriptors were obtained using the melodic segmentation approach of Cambouropoulos (1997), which indicates the probability of each note being at a phrase boundary. Probability values were used to decide if the note was a *boundary note*, annotated as either *initial (i)* or *ending (e)*. Non-boundary notes were annotated as *middle (m)*. The phrase descriptor was introduced based on the hypothesis that boundary notes (i.e. initial or ending phrase notes) are more prone to be ornamented than middle notes. *Note to key* and *note to chord* descriptors are intended to capture harmonic analysis information, as they refer to the interval of a particular note with respect to the key and to the chord root, respectively. *Key* and *mode* refer to the key signature of the song (e.g. key: *C*, mode: major). Mode is a binary descriptor (major or minor), whereas key is represented numerically in the circle of fifths (e.g. $B\flat = -1$, $C = 0$, $F = 1$, etc.). However, for some calculations (e.g. note to key in Table 3) a linear representation of the notes (e.g. $C = 0$, $C\sharp/D\flat = 1$, $D = 2$, etc.) is used for key. Also, it is worth noticing that the key descriptor may have 13 possible values as the extreme values (-6 and 6) corresponding to enharmonic tonalities ($G\flat$ and $F\sharp$). The descriptor *Is chord note* was calculated using the chord type description of Table 1 in which each of the notes of the chord is shown using the aforementioned linear note representation. If a note corresponds to any of the notes included in the chord type description it is labeled *yes*.

Table 1. Chord description list.

Chord type	Intervals
major	0 4 7
m (minor)	0 3 7
2 (sus2)	0 2 7
sus (4)	0 5 7
Maj7	0 4 7 11
6th	0 4 7 9
m7	0 3 7 10
m6	0 3 7 9
mMaj7	0 3 7 11
m7♭5	0 3 6 10
dim	0 3 6 9
7th	0 4 7 10
7♯5	0 4 8 10
7♭5	0 4 6 10
7sus	0 5 7 10
Maj9	0 2 4 7 11
m9	0 2 3 7 11
6/9	0 2 4 7 9
m6/9	0 2 3 7 9
9th	0 2 4 7 10
7♭9	0 1 4 7 10
7♯9	0 3 4 7 10
13	0 2 4 7 9 10
7♭9♭13	0 1 4 7 8 10
7alt	0 1 3 4 6 8 10

Other relevant descriptors used for this work include categorization based on the *implication–realization* (I-R) model of Narmour (1992). Grachten (2006) parses the melodies and obtains for each note the label of the I-R structure to which it belongs. The concept of closure was based on metrical position and duration. The basic Narmour structures (P, D, R, and ID) and their derivatives (VR, IR, VP, and IP) are represented in Figure 2.

The *metrical strength* concept refers to the rhythmic position of the note inside the bar (Cooper and Meyer 1960). Four levels of metrical strength were used to label notes in three common time signatures, depending on the beat at which the note occurs, as shown in Table 2.

The complete list of the 30 descriptors used for this study and its definition is summarized in Table 3.

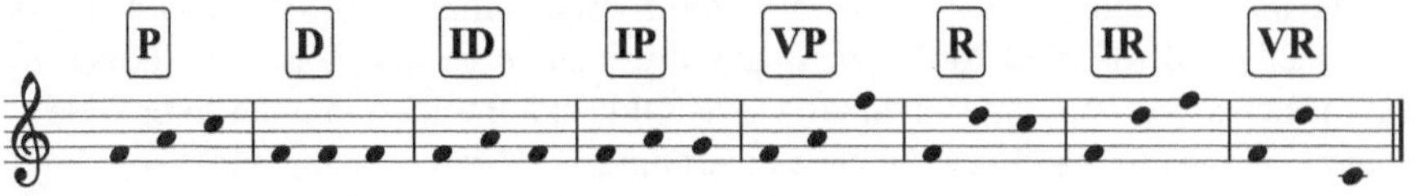

Figure 2. Basic Narmour structures P, D, R, and ID, and their derivatives VR, IR, VP, and IP.

Table 2. Metrical strength at beat occurrence, for different time signatures.

Time signature	Very strong	Strong	Weak	Very weak
4/4	Beat 1	Beat 3	Beats 2 and 4	Other
3/4	Beat 1	None	Beats 2 and 3	Other
6/8	Beat 1	Beat 2.5	Beats 1.5, 2, 3, and 3.5	Other

Table 3. Features extracted from music scores – in the fifth row, in the column headed 'Formula,' '*bpb*' means *beats per bar* according to the time signature.

	Descriptor	Abbreviation	Units	Formula	Range
Nominal	Duration	ds_n	Seconds	ds_0	$[0, +\infty]$
	Duration	db_n	Beats	db_0	$[0, +\infty]$
	Onset	ons_n	Seconds	os_0	$[0, +\infty]$
	Onset	onb_n	Beats	ob_0	$[0, +\infty]$
	Onset in bar	obm_n	Beats	$ob_0\%bpb$	$[0, +bpb]$
	Pitch	p_n	Semitones	p_0	$[1, 127]$
	Chroma	ch_n	Semitones	$p_0\%12$	$[0, 11]$
	Energy	v_n	MIDI vel	v_0	$[1, 127]$
Neighbor	Prev. duration	pds_n	Seconds	ds_{-1}	$[0, +\infty]$
	Prev. duration	pdb_n	Beats	db_{-1}	$[0, +\infty]$
	Next duration	nds_n	Seconds	ds_1	$[0, +\infty]$
	Next duration	ndb_n	Beats	db_1	$[0, +\infty]$
	Prev. interval	$pint_n$	Semitones	$p_{-1} - p_0$	$[-60, 60]$
	Next interval	$nint_n$	Semitones	$p_1 - p_0$	$[-60, 60]$
	Prev. inter-onset dist.	$piod_n$	Seconds	$os_0 - os_{-1}$	$[0, +\infty]$
	Next. inter-onset dist.	$piod_n$	Seconds	$os_1 - os_0$	$[0, +\infty]$
	Narmour	$nar1_n$ $nar2_n$ $nar3_n$	Label	$nar(p_{-1}, p_0, p_1)$ $nar(p_{-2}, p_{-1}, p_0)$ $nar(p_0, p_1, p_2)$	{P, D, R, ID, (P), (R), (ID), VR, IR, VP, IP, (VR), (IR), (VP), (IP), dyadic,' monadic, none}
Context	Measure	m_n	Bars	m_0	$[0, +\infty]$
	Tempo	t_n	Bpm	t_0	$[30, 260]$
	Key	k_n	Semitones	k_0	$[-6, 6]$
	Mode	mod_n	Label	mod_0	{major, minor}
	Note to key	$n2k_n$	Semitones	$ch_0 - k_0$ (linear)	$[0, 11]$
	Chord root	chr_n	Semitones	chr_0	$[0, 11]$
	Chord type	cht_n	Label	cht_0	{+, 6, 7, 7♯11, 7♯5, 7♯9, 7alt, 7♭5, 7♭9, Maj7, dim, dim7, m, m6, m7, m7♭5, major}
	Note to chord	$n2ch_n$	Semitones	$ch_0 - chr_0$	$[0, 11]$
	Is chord note	$ichn_n$	Boolean	$isChNote(chr_0, cht_0, ch_0)$	{true, false}
	Metrical strength	mtr_i	Label	$metStr_0$	{Very strong, Strong, Weak, Very weak}
	Phrase	ph_n	Label	$phrase_0$	{initial, middle, final}

3.2. *Audio analysis*

The audio of the performed pieces was recorded from the raw signal of an electric guitar. The guitarist was instructed not to strum chords or play more than one note at a time. The guitarist recorded the pieces while playing along with prerecorded commercial accompaniment backing tracks (Kennedy and Kernfeld 2002). We opted to use audio backing tracks performed by professional musicians, as opposed to synthesized MIDI backing tracks, in order to provide a more natural and ecologically valid performance environment. However, using audio backing tracks required a preprocessing beat tracking task. Each piece's section was recorded once (i.e. no repetitions or solos were recorded), For instance, for a piece consisting of sections *AABB*, only sections *A* and *B* were considered.

3.2.1. *Melodic transcription*

The monophonic audio signal recorded from the guitar was parsed to automatically obtain a *MIDI* type transcription of the notes performed by the guitarist, based on the previous work of Bantula, Giraldo, and Ramírez (2014). This representation includes the pitch, onset, duration, and energy of each note. For doing this, the audio signal was segmented based on the pitch and energy profiles obtained with the YIN algorithm (De Cheveigné and Kawahara 2002). Each segment represents a note with its corresponding information on pitch, onset, and offset (therefore duration). To minimize transcription errors, the resulting segments (notes) were filtered using heuristic rules based on human perceptual thresholds for minimum note duration and minimum note gaps. Also, rules to detect octave errors or unusual note intervals were used. To obtain temporal information on the recordings, beat tracking (Zapata et al. 2012) was used, as there was uncertainty concerning the use of a metronome in the recordings of the accompaniment backing tracks. After manual correction of beat tracking, the onset and duration information on each note was adjusted to the beat grid detected for each piece.

3.3. *Score to performance alignment*

Score to performance alignment was performed to correlate each performed note with its respective *parent* note in the score as depicted in Figure 3. This procedure was carried out following the approach of Giraldo and Ramírez (2015d), in which DTW techniques were used to match performance and score note sequences. A similarity cost function was designed based on pitch, duration, onset, and phrase onset/offset deviations.

Figure 3. Score to performance alignment. Fragment of *Yesterdays* (J. Kern) as performed by Wes Montgomery.

Phrase onset and offset deviation were introduced to force the algorithm to map all the notes of a particular short ornament phrase (*lick*) to one parent note in the score. We assumed that a group of notes conforming a *lick* are played *legato*. Therefore, the performed sequence is segmented in phrases in which the time gap between consecutive notes is less than 50 ms. This threshold was chosen based on human time perception studies (Woodrow 1951).

Each note from the score and the corresponding performed sequence is represented by a five position *cost vector* as

$$cs = (p(i), ds(i), ons(i), ons(i), ofs(i)) \tag{1}$$

and

$$cp = (p(j), ds(j), ons(j), ph_{ons}(j), ph_{ofs}(j)), \tag{2}$$

respectively, where *cs* is the *score cost vector* and *cp* is the *performance cost vector*. Index *i* refers to a note position in the score sequence, and *j* refers to a note position at the performed sequence. The onset of the first note of the lick phrase in which the *j*th note of the performance sequence occurs is represented by $ph_{ons}(j)$. Similarly $ph_{ofs}(j)$ refers to the offset of the last note of the lick phrase in which the *j*th note of the performance sequence occurs.

The total cost is calculated using the *Euclidean distance* as follows:

$$cost(i,j) = \sqrt{\sum_{n=1}^{5} (cs(n)_i - cp(n)_j)^2}\,. \tag{3}$$

Notice that in equation (3) phrase onset and offset deviations are calculated when n equals four and five.

Finally, we apply DTW: a similarity matrix $H_{(m \times n)}$ is defined in which m is the length of the performed sequence of notes and n is the length of the sequence of score notes. Each cell of the matrix H is calculated as follows:

$$H_{i,j} = cost + min(H_{i-1,j}, H_{i,j-1}, H_{i-1,j-1}) \tag{4}$$

where *min* is a function that returns the minimum value of the preceding cells (up, left, and up-left diagonal). The matrix H is indexed by the note position of the score sequence and the note position of the performance sequence.

A backtrack path is obtained by finding the lowest cost calculated in the similarity matrix. Starting from the last score/performance note cell, the cell with the minimum cost at positions $H_{(i-1)}$, $H_{(i,j-1)}$, and $H_{(i-1,j-1)}$ is stored in a backtrack path array. The process iterates until indexes arrive to the first position of the matrix, assigning each note in the performance to a parent note in the score.

Figure 4 presents an example of the resulting similarity matrix obtained for one of the recorded songs. The x-axis corresponds to the sequence of notes of the score and the y-axis corresponds to the sequence of performed notes. The cost of correspondence between all possible pairs of notes is depicted darker for the highest cost (less similar) and lighter for the lowest cost (most similar).

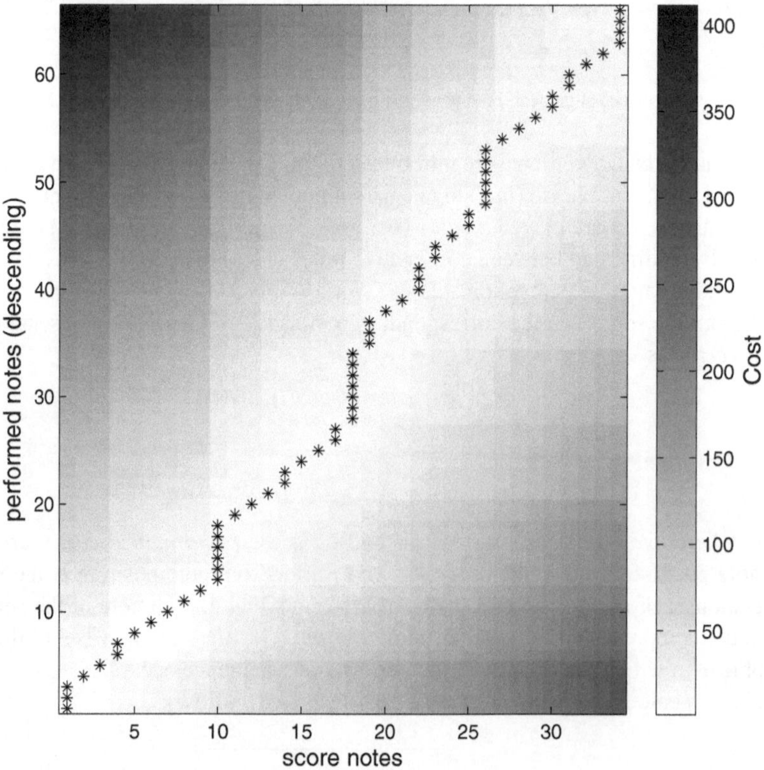

Figure 4. Similarity matrix of performed notes and score notes. Dots indicate alignment path between score and performance notes.

The dots on the graph show the backtrack path (or optimal path) found for alignment. Diagonal lines represent notes which were not ornamented, as the correspondence from the performance notes to the parent score notes is one to one. On the contrary, vertical lines represent ornaments, as two or more performed notes correspond to one parent note in the score.

Because there are no concrete rules to map performance notes to *parent* score notes, our alignment algorithm was evaluated by comparing its output with the level of agreement between five human experts who were asked to align performance and score note sequences manually. Accuracy of the system was estimated by quantifying how much each note pair produced by the algorithm agreed with the human experts, using penalty factors for high, medium, and low agreement. The results of the evaluation showed that the performance of our approach was comparable with that of the human annotators. Details of these evaluations can be found in Giraldo and Ramírez (2015d).

3.4. *Performance actions*

After alignment, score notes which were mapped to only one performance note were labeled as *non-ornamented*, whereas score notes mapped to several performance notes (or which were omitted in the performance) were mapped as *ornamented.* Performance actions were calculated

Table 4. Expressive performance actions calculation.

PA	Abbreviation	Units	Formula	Range
Ornamentation	Orn_n	Boolean	$ornament(N_n)$	{yes, no}
Duration ratio	Dr_n	Percentage	$\frac{db_j}{db_i} * 100$	$[0, +\infty]$
Onset deviation	Od_n	Beats	$ob_j - ob_i$	$[0, +\infty]$
Energy ratio	Er_n	Percentage	$\frac{v_j}{70} * 100$	$[0, +\infty]$

Table 5. Example of ornament annotation for the music excerpt of Figure 3.

Score note index (*i*)	Performance note index (*j*)	Pitch deviation (semitones)	Onset deviation (beats)	Duration ratio (beat fraction)
1	1	−1	−1/2	1/16
1	2	0	0	2/3
2	3	−3	−1/2	1/2
2	4	0	0	1/2
4	6	0	−1/2	1/16
4	7	0	1/2	1/16
4	8	−1	3/2	1/16
4	9	0	2	1/16
5	10	−3	−1/2	1/2
5	11	0	0	1
6	12	1	−1/2	1/8
6	13	0	0	1/8
6	14	−2	1/2	1/8
6	15	0	1	1/8
6	16	0	3/2	1/8

for each score note, as defined in Table 4, by measuring the deviations in onset, energy, and duration. Again, indexes *i* and *j* refer to the note position at the score and the performance sequence, respectively.

3.5. *Database construction*

The data collected was organized, storing each note descriptors along with its corresponding *performance action*. The *pitch*, *duration*, *onset*, and *energy* deviations of each ornament note with respect to the score *parent* note were annotated as shown in Table 5.

4. Expressive performance modeling

Several machine learning algorithms – i.e. *artificial neural networks* (ANNs), *decision trees* (DTs), *support vector machines* (SVMs), and *k-nearest neighbor* (k-NN) – were applied to predict the ornaments introduced by the musician when performing a musical piece. The accuracy of the resulting classifiers was compared with the *baseline classifier*, i.e. the classifier which always selects the most common class. Timing, onset, and energy performance actions were modeled by applying several regression machine learning methods (i.e. ANNs, *regression trees* (RTs), SVMs, and k-NN).

Based on their accuracy, we chose the best performance model and feature set to predict the different performance actions. Each piece (used as test set) was in turn predicted based on the models obtained with the remaining pieces (used as training set) and synthesized using a concatenative synthesis approach.

4.1. *Algorithm comparison*

In this study we compared four classification algorithms for ornament prediction and four regression algorithms for duration ratio, onset deviation, and energy ratio. We used the implementation of the machine learning algorithms provided by the WEKA library (Hall et al. 2009). We applied k-NN with $k=1$, SVM with linear kernel, ANN consisting of a fully-connected multi-layer neural network with one hidden layer, and DT/RT with post pruning.

A paired T-test with a significance value of 0.05 was performed for each algorithm for the ornamentation classification task, over all the data set with 10 runs of 10-fold cross validation scheme. Experiments results are presented in Table 8 and will be commented in Section 6.

4.2. *Feature selection*

Both filter and wrapper feature selection methods were applied. Filter methods use a proxy measure (e.g. information gain) to score features, whereas wrappers make use of predictive models to score feature subsets. Features were filtered and ranked by information gain values, and a wrapper with greedy search and decision trees accuracy evaluation was used to select optimal feature subsets. We used the implementation of these methods provided by WEKA library (Hall et al. 2009). Selected features are shown in Table 6, and will be commented upon in Section 6.

Learning curves on the number of features, as well as on the number of instances, were obtained to measure the learning rate of each of the algorithms. The selection of the model was based on the evaluation obtained with these performance measures.

Table 6. Features selected using the Filter and Wrapper (with J48 classifier) methods.

Info-Gain + Ranker	Wrapper + Greedy
Duration (sec)	Duration beat
Duration (beat)	Prev. duration (sec)
Phrase	Tempo
Prev. duration (sec)	Phrase
Onset in bar	Velocity
Metrical strength	Onset beat
Prev. duration (beat)	Duration (sec)
Next duration (beat)	Prev. duration (beat)
Next duration (sec)	Next duration (beat)
Narmour 1	Narmour 3
Tempo	Metrical strength
Chord type	Onset in bar
Prev. interval	Narmour 1
Next interval	Prev. interval
Narmour 2	Narmour 2
Narmour 3	Is chord note
Is chord note	Onset (sec)
Mode	Key
keyMode	Chord root
	Next duration (sec)
	Pitch
	Note to key
	Note to chord
	Measure
	Next interval
	Chroma
	Chord type

5. Synthesis

Predicted pieces were created in both MIDI and audio formats. A concatenative synthesis approach was used to generate the audio pieces. This process consists of linking note audio samples from real performances to render a synthesis of a musical piece. The use of this approach was possible because we had monophonic performance audio data from which onset and offset information was extracted based on energy and pitch information as described in Section 3.2.1. Therefore, it was possible to segment the audio signal into individual notes, and furthermore to obtain complete audio segments of ornaments.

Similarly to the evaluation of the different machine learning algorithms, for synthesis we followed a leave-one-piece-out scheme in which, on each fold, the notes of one piece were used as test set, whereas the notes of the remaining 26 songs were used as training set.

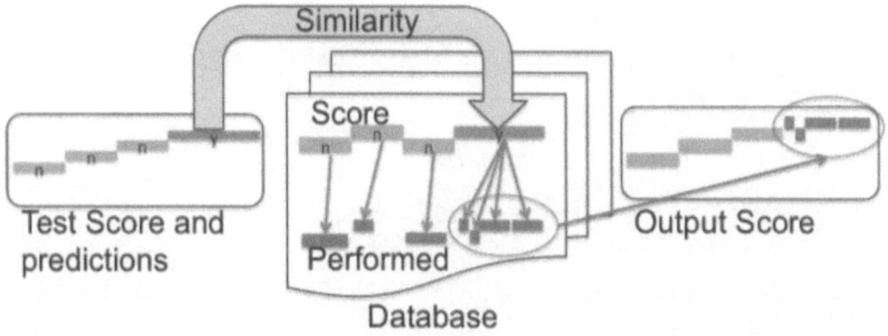

Figure 5. Concatenative synthesis approach.

5.1. *Note concatenation*

The note concatenation process is divided into three different stages as depicted in Figure 5.

- *Sample retrieval* For each note predicted to be ornamented, the k-NN algorithm, using a *Euclidean* distance similarity function based on note description, was applied to find the most suitable ornamentation in the database (see Section 3.5). This was done by searching for the most suitable ornament in the songs in the training set (Section 4.2).
- *Sample transformation* For each note classified as ornamented, transformations in duration, onset, energy, and pitch (in the case of ornaments) were performed based on the deviations stored in the database, as seen in Figures 6(a) and 6(b). For audio sample transformation we used the time and pitch scaling approaches of Serra (1997). Notes classified as not ornamented were simply transformed as predicted by the duration, onset, and energy models.
- *Sample concatenation* Retrieved samples were concatenated based on final onset and duration information after transformation. The tempo of the score being predicted (in BPM), was imposed on all the retrieved notes.

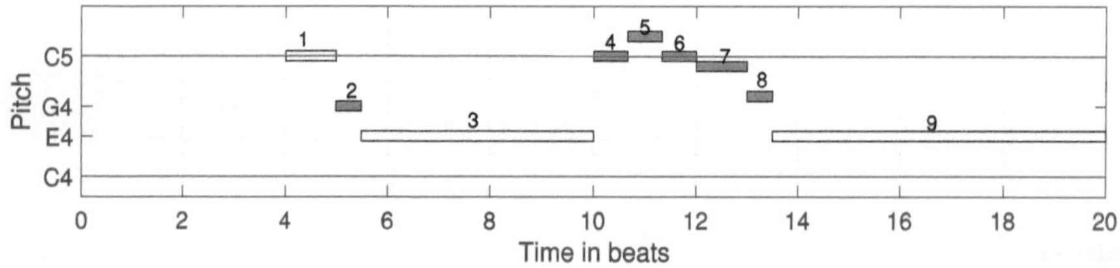

(a) Piano roll of a score (*All of me*). Gray and white boxes represent notes predicted as *not ornamented* and *ornamented*, respectively. The transformation for note 3 is explained in Figures (b) and (c).

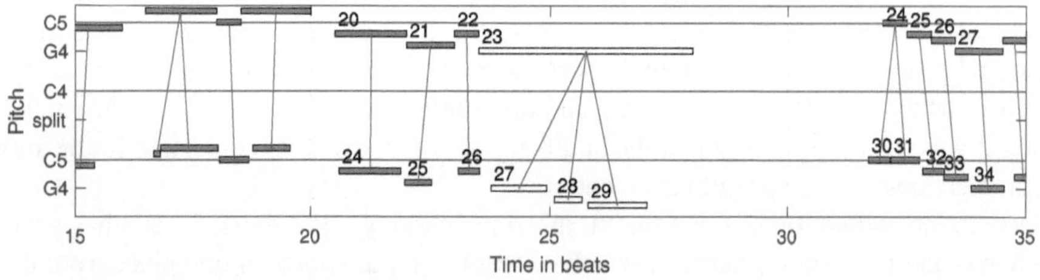

(b) Piano roll of the most similar note found. Top sequence represents the score in which the closest note (note 23) was found (*Satin Doll*). Bottom sequence represents the performance of the score. Vertical lines show note correspondence between score and performance.

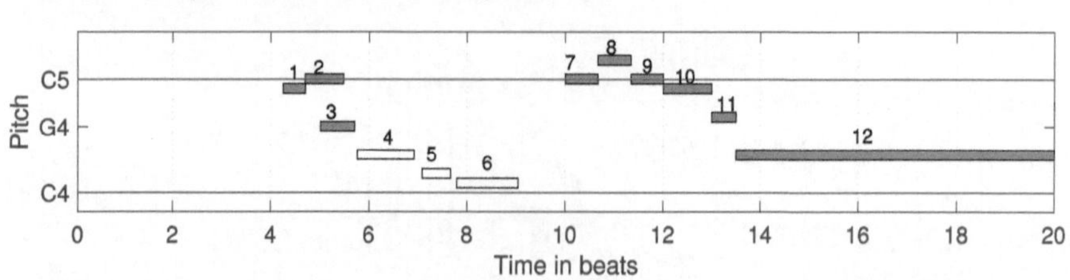

(c) Piano roll of a partial predicted score (*All of me*). Note number 3 of Figure (a) has been replaced by notes 27, 28, and 29 of Figure (b), obtaining notes 4, 5, and 6.

Figure 6. Sample retrieval and concatenation example.

6. Results

6.1. *Feature selection*

The most relevant features found using the two selection methods described in Section 4.2 are shown in Table 6. The average correctly classified instances percentage (C.C.I.%) obtained using the features selected by the information gain filtering and the greedy search (decision trees) wrapper methods were 78.12 and 78.60%, respectively (F-measures of 0.857 and 0.866, respectively). Given that both measures are similar, i.e. not significantly different, the smallest subset was chosen.

In Figure 7, the accuracy on increasing the number of features based on the information gain ranking (explained in Section 4.2) is presented for each of the algorithms used (SVM, ANN, DT). From the curves it can be seen that the subset with the first three features contains sufficient information, as additional features do not add significant accuracy improvement. SVM exhibits better accuracy on the cross validation scheme, and less over-fitting based on the difference between *cross validation* (CV) and *training set* (TS) accuracy curves.

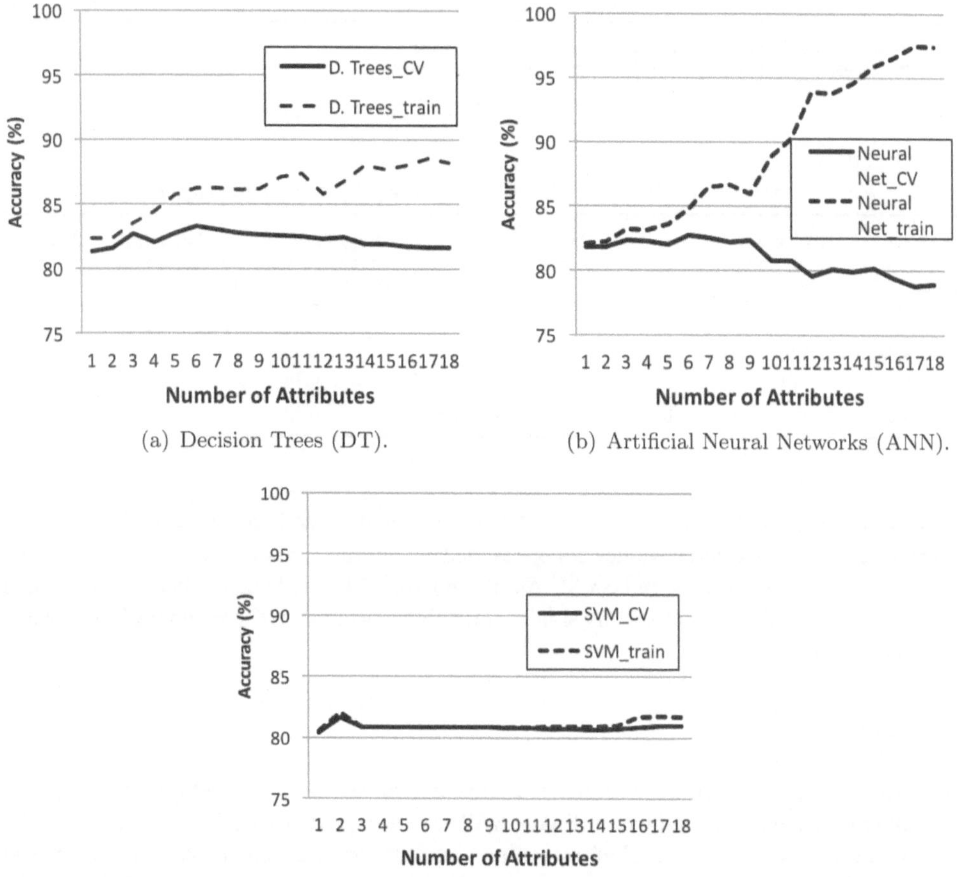

(a) Decision Trees (DT). (b) Artificial Neural Networks (ANN). (c) Support Vector Machines (SVM).

Figure 7. Accuracies on increasing the number of attributes.

6.2. *Quantitative evaluation*

6.2.1. *Algorithm comparison results*

For the (ornament) classification problem we compared each of the algorithms (SVM, DT, ANN, and k-NN) with the baseline classifier (i.e. the majority class classifier) following the procedure explained in Section 4.1. From Table 7 it can be seen that all the algorithms present a statistically significant improvement, except k-NN. Given the accuracy results, we apply the ornamentation prediction model induced by the DT algorithm to determine whether a note is to be ornamented or not. We discarded the use of k-NN for this task due to its low accuracy, which led to larger mis-classifications of ornamented and not ornamented notes.

For the regression problems (duration, onset, and energy prediction) we applied *regression trees*, *SVM*, *neural networks*, and *k-NN*, and obtained the correlation coefficient values shown in Table 8. *Onset deviation* has the highest correlation coefficient, close to 0.5.

Table 7. *Correctly classified instances* (CCIs) (%) comparison with paired T-test for classification task.

Dataset	Baseline classifier (CCIs) (%)	Instance base learner (CCIs) (%)	Design trees (CCIs) (%)		SVM (CCIs) (%)		Neural Network (CCIs) (%)	
Ornament	72.74	70.58	78.68	∘⋄	77.64	∘⋄	76.60	∘⋄

∘/• Statistically significant improvement/degradation against 'Baseline classifier.'
⋄/⋆ Statistically significant improvement/degradation against 'Instance base learner.'

Table 8. *Pearson correlation coefficient* (PCC) and *explained variance* (R^2) for the regression task.

	k-NN			Reg. Trees			Reg. SVM		ANN		PA_{mean}	
Dataset	PCC	R^2		PCC	R^2		PCC	R^2	PCC	R^2	PCC	R^2
Duration ratio	0.20	0.04		0.25	0.06		0.17	0.03	0.19	0.04	0.20	0.04
Energy ratio	0.19	0.04		0.37	0.14		0.38	0.14	0.37	0.14	0.33	0.11
Onset deviation	0.41	0.17		0.51	0.26		0.43	0.18	0.44	0.19	0.45	0.20
$Algorithm_{mean}$	0.25	0.08	0.38	0.15	0.33	0.12	0.33	0.12				

For ornamentation classification using k-NN we explored several values for k ($1 \leq k \leq 10$). However, all of the explored values for k resulted in inferior classification accuracies when compared with decision trees and SVM. As in the case of $k = 1$, both the decision trees and SVM classifiers resulted in statistically significantly higher accuracies (based on the T-test) when compared with the classifiers for $2 \leq k \leq 10$.

6.3. *Learning curves*

The learning curves of accuracy improvement, for both cross validation and training sets, over the number of instances are shown in (Figure 8). The learning curves were used to measure the learning rate and estimate the level of overfitting. Data subsets of different sizes (in steps of 100 randomly selected instances) were considered and evaluated using 10-fold cross validation. In general, for the three models, it can be seen that the accuracy on CV tends to have no significant improvement above 600 instances.

Overfitting can be correlated with the difference between the accuracies of CV and TS, wherein a high difference means higher levels of overfitting. In this sense, in Figure 8(c), SVM shows a high tendency for overfitting, but seems to improve slowly over the number of instances. On the other hand, in Figures 8(a) and 8(b), ANN and DT seem to improve overfitting between 700 and 1100 instances. This could mean that adding more instances may slightly improve the accuracy of both CV and TS for the three models, and may slightly improve overfitting for SVM, but this may not be the case for ANN and DT.

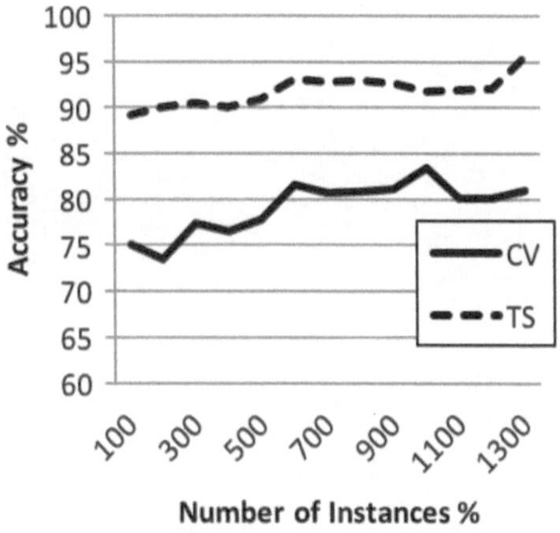

(a) Decision Trees (DT).

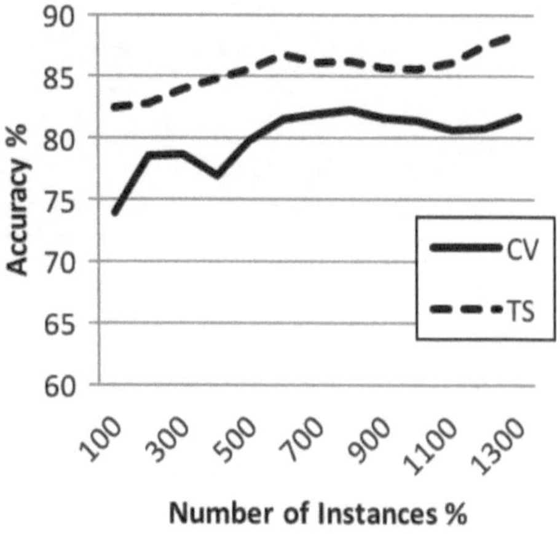

(b) Artificial Neural Networks (ANN).

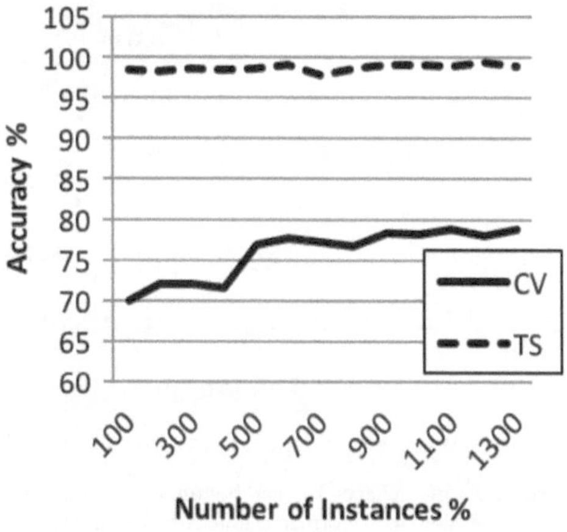

(c) Support Vector Machines (SVM).

Figure 8. Accuracies on increasing the number of instances.

6.4. *Obtained pieces*

Figure 9 shows a MIDI piano roll of an example piece performed by a professional musician and the predicted performance obtained by the system, using a decision trees classifier. It can be noticed how the predicted piano roll follows a similar melodic structure as the one performed by the musician. For instance, for the score notes predicted correctly as ornamented (true positives), notes 1, 10, and 34 in Figure 9(a) (top sequence), the system finds ornaments of similar duration, offset, and number of notes as the musician's performance. Also, score notes 3 and 9 of Figure 9(b) (false positives), are ornamented similarly as score notes 18 and 26 (Figure 9(a)), which are in a similar melodic context.

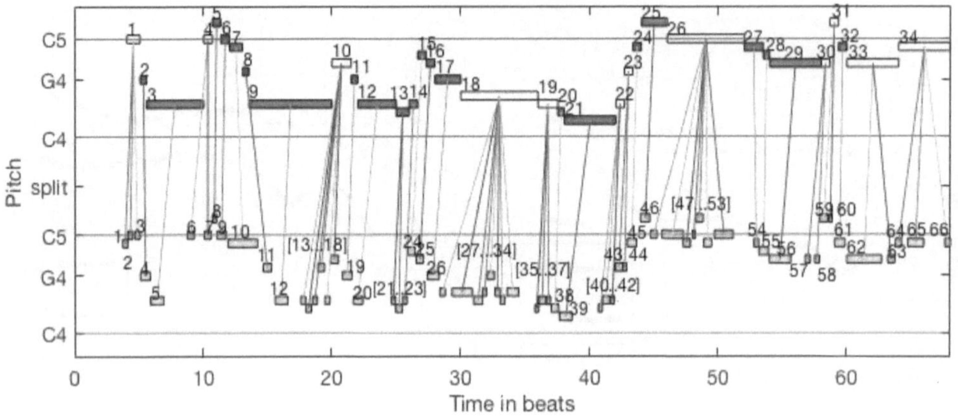

(a) Score to musician performance correspondence.

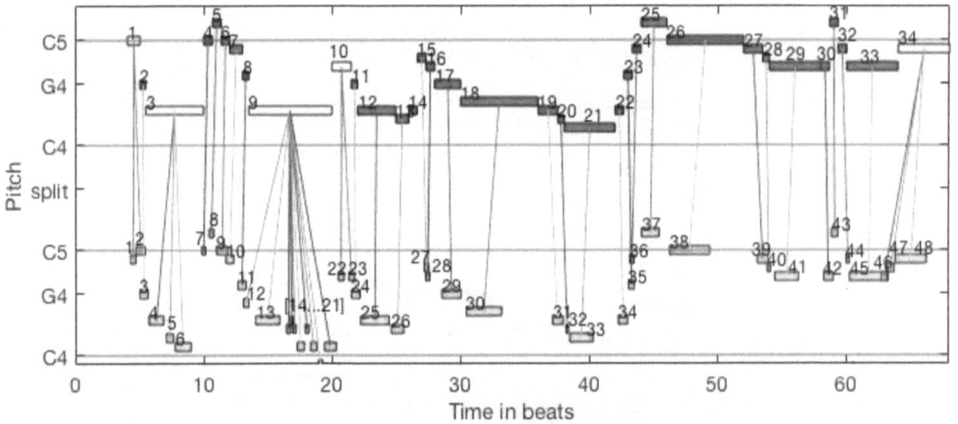

(b) Score to predicted performance correspondence.

Figure 9. Musician versus predicted performance. Top and bottom sequences represent score and performance piano roll, respectively. Vertical lines indicate score to performance note correspondence. Gray and white boxes represent notes predicted as *not ornamented* and *ornamented*, respectively.

6.5. *Duration and energy ratio curves*

Duration and energy deviation ratio measured in the musician performance and predicted by the system for one example piece (*All of me*) are compared in Figures 10(a) and 10(b), respectively. We obtained similar results for the other pieces in the data set. Similarity between the contour of the curves indicates that the deviations predicted by the system are coherent with the ones performed by the musician.

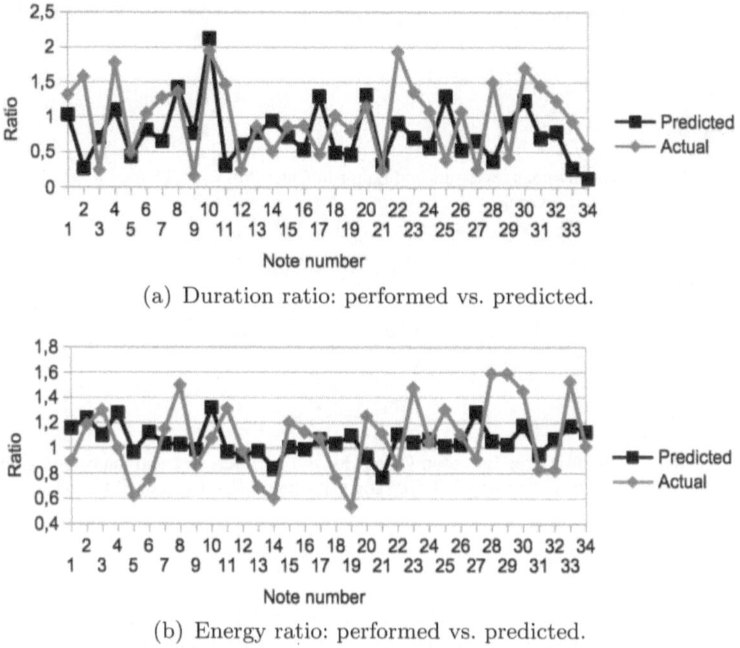

Figure 10. Performed versus predicted duration and energy ratio example for *All of me*. Gray and black lines represent performed and predicted ratios, respectively, for each note in the score.

6.6. *Musical samples*

Musical examples of the automatically generated ornamented pieces can be found in the Online Supplement (see the unnumbered section directly before the references list at the end of this article). The rendered audio of the *Yesterdays* music piece generated by the system (as test piece) has been included in this site.

7. Conclusions

In this article we have presented a machine learning approach for expressive performance (ornament, duration, onset, and energy) prediction and synthesis in jazz guitar music. We used a data set of 27 recordings performed by a professional jazz guitarist, and extracted a set of descriptors from the music scores and a symbolic representation from the audio recordings. In order to map performed notes to parent score notes we have automatically aligned performance to score data. Based on this alignment we obtained performance actions, calculated as deviations of the performance from the score. We created an ornaments database including the information on the ornamented notes performed by the musician. We have compared four learning algorithms to create models for ornamentation, based on performance measures, using a significance paired T-test. Feature selection techniques were employed to select the best feature subset for ornament modeling. For synthesis purposes, instance based learning was used to retrieve the most suitable ornament from the ornamentation database. A concatenative synthesis approach was used to generate expressive performances of new pieces – i.e. pieces not in the training set – automatically. A subjective perceptual evaluation based on listening tests is beyond the scope of this article. As future work, we plan to evaluate the performances generated by the system by computing the alignment distance between the system and the target performance.

Acknowledgments

We would like to thank the two anonymous reviewers and Editor-in-Chief Professor Thomas Fiore for their constructive comments on previous versions of this article.

Disclosure statement

No potential conflict of interest was declared by the authors.

Funding

This project has received funding from: the European Union Horizon 2020 research and innovation programme [grant agreement No 688269]; the Spanish TIN project TIMUL [grant agreement TIN2013-48152-C2-2-R].

Supplemental online material

Supplemental online material for this article can be accessed at doi:10.1080/17459737.2016.1207814 and https://soundcloud.com/machine-learning-and-jazz. In the Online Supplement we present an example of the generated pieces. The Online Supplement consists of a pdf description file and three audio files. The three audio files are an inexpressive (mechanical) rendering of the score, a recorded performance of the musician, and the rendered performance predicted by the system.

References

Arcos, Josep Lluís, Ramon López De Mántaras, and Xavier Serra. 1998. "SaxEx: A Case-Based Reasoning System for Generating Expressive Musical Performances." *Journal of New Music Research* 27 (3): 194–210.

Bantula, Helena, Sergio Giraldo, and Rafael Ramírez. 2014. "A Rule-Based System to Transcribe Guitar Melodies." In *Proceedings of the 11th International Conference on Machine Learning and Music (MML 2014)*, 28 November 2014, Barcelona, Spain, 6–7.

Bresin, Roberto. 1998. "Artificial Neural Networks Based Models for Automatic Performance of Musical Scores." *Journal of New Music Research* 27 (3): 239–270.

Bresin, Roberto, and Anders Friberg. 2000. "Emotional Coloring of Computer-Controlled Music Performances." *Computer Music Journal* 24 (4): 44–63.

Cambouropoulos, Emilios. 1997. "Musical Rhythm: A Formal Model for Determining Local Boundaries, Accents and Metre in a Melodic Surface." In *From Rhythm to Expectation*. Volume III of *Music, Gestalt, and Computing: Studies in Cognitive and Systematic Musicology*. Volume 1317 of Lecture Notes in Computer Science Series, 277–293. Berlin: Springer. doi:10.1007/BFb0034121.

Camurri, Antonio, Roberto Dillon, and Alberto Saron. 2000. "An Experiment on Analysis and Synthesis of Musical Expressivity." *Proceedings of the 13th Colloquium on Musical Informatics (XIII CIM)*, 2–5 September 2000, L'Aquila: Italy. ftp://ftp.infomus.org/pub/Publications/2000/CIM2000CDS.PDF.

Cooper, Grosvenor, and Leonard B. Meyer. 1960. *The Rhythmic Structure of Music*. Chicago, IL: University of Chicago Press. http://www.press.uchicago.edu/ucp/books/book/chicago/R/bo24515499.html.

De Cheveigné, Alain, and Hideki Kawahara. 2002. "YIN, a Fundamental Frequency Estimator for Speech and Music." *The Journal of the Acoustical Society of America* 111 (4): 1917–1930.

Friberg, Anders. 2006. "pDM: An Expressive Sequencer with Real-Time Control of the KTH Music-Performance Rules." *Computer Music Journal* 30 (1): 37–48.

Friberg, Anders, Roberto Bresin, and Johan Sundberg. 2006. "Overview of the KTH Rule System for Musical Performance." *Advances in Cognitive Psychology* 2 (23): 145–161.

Gabrielsson, Alf. 1999. "The Performance of Music." In *The Psychology of Music*, edited by Diana Deutsch, Cognition and Perception Series, 2nd ed., 501–602. San Diego, CA: Academic Press.

Gabrielsson, Alf. 2003. "Music Performance Research at the Millennium." *Psychology of Music* 31 (3): 221–272.
Giraldo, Sergio. 2012. "Modeling Embellishment, Duration, and Energy Expressive Transformations in Jazz Guitar." Master's thesis, Universidad Pompeu Fabram, Barcelona, Spain.
Giraldo, Sergio, and Rafael Ramírez. 2014. "Optimizing Melodic Extraction Algorithm for Jazz Guitar Recordings Using Genetic Algorithms." In *Joint 42nd International Computer Music Conference and 13rd Sound & Music Computing Conference (ICMC-SMC 2014)*, 22–26 October 2014, Athens, Greece, 25–27.
Giraldo, Sergio, and Rafael Ramírez. 2015a. "Computational Generation and Synthesis of Jazz Guitar Ornaments using Machine Learning Modeling." In *Proceedings of the 11th International Conference on Machine Learning and Music (MML 2014)*, August 2015, Vancouver, Canada, 10–12.
Giraldo, Sergio, and Rafael Ramírez. 2015b. "Computational Modeling and Synthesis of Timing, Dynamics and Ornamentation in Jazz Guitar Music." In *11th International Symposium on Computer Music Interdisciplinary Research (CMMR 2015)*, Plymouth, UK, 806–814.
Giraldo, Sergio, and Rafael RafaelRamírez. 2015c. "Computational Modelling of Ornamentation in Jazz Guitar Music." In *International Symposium in Performance Science*, 2 September 2015, Kyoto, Japan, 150–151.
Giraldo, Sergio, and Rafael Ramírez. 2015d. "Performance to Score Sequence Matching for Automatic Ornament Detection in Jazz Music." In *International Conference of New Music Concepts (ICMNC 2015)*, Treviso, Italy, 8.
Goebl, Werner, Simon Dixon, Giovanni DePoli, Anders Friberg, Roberto Bresin, and Gerhard Widmer. 2008. "Sense in Expressive Music Performance: Data Acquisition, Computational Studies, and Models." Chapter 5 in *Sound to Sense – Sense to Sound: A State of the Art in Sound and Music Computing*, 195–242. Berlin: Logos. http://iwk.mdw.ac.at/goebl/papers/Goebl-etal-2008-Sense-in-Expressive-Performance.pdf.
Gómez, Francisco, Aggelos Pikrakis, Joaquín Mora, Juan Manuel Díaz-Bánez, Emilia Gómez, and Francisco Escobar. 2011. "Automatic Detection of Ornamentation in Flamenco." In *Fourth International Workshop on Machine Learning and Music (MML 2011)*, 20–22.
Grachten, Maarten. 2006. "Expressivity-Aware Tempo Transformations of Music Performances Using Case Based Reasoning." PhD thesis, Universidad Pompeu Fabra: Barcelona, Spain.
Grindlay, Graham Charles. 2005. "Modeling Expressive Musical Performance with Hidden Markov Models." Master's thesis: University of California Santa Cruz. http://www.ee.columbia.edu/~grindlay/pubs/UCSC_MS_thesis_2005.pdf.
Hall, Mark, Eibe Frank, Geoffrey Holmes, Bernhard Pfahringer, Peter Reutemann, and Ian H. Witten. 2009. "The WEKA Data Mining Software: An Update." *ACM SIGKDD Explorations Newsletter* 11 (1): 10–18. http://www.cms.waikato.ac.nz/~ml/publications/2009/weka_update.pdf.
Johnson, Margaret L. 1991. "Toward an Expert System for Expressive Musical Performance." *Computer* 24 (7): 30–34.
Kennedy, Gary, and Barry Kernfeld. 2002. "Aebersold, Jamey." In *The New Grove Dictionary of Jazz.* Volume 1. 2nd ed. 16–17. New York: Grove's Dictionaries.
Kirke, Alexis, and Eduardo R. Miranda. 2013. "An Overview of Computer Systems for Expressive Music Performance." In *Guide to Computing for Expressive Music Performance*, 1–47. London: Springer.
Miranda, Eduardo R., Alexis Kirke, and Qijun Zhang. 2010. "Artificial Evolution of Expressive Performance of Music: An Imitative Multi-Agent Systems Approach." *Computer Music Journal* 34 (1): 80–96.
Narmour, Eugene. 1992. *The Analysis and Cognition of Melodic Complexity: The Implication–Realization Model.* Chicago, IL: University of Chicago Press.
Palmer, Caroline. 1997. "Music Performance." *Annual Review of Psychology* 48 (1): 115–138.
Perez, Alfonso, Esteban Maestre, Stefan Kersten, and Rafael Ramírez. 2008. "Expressive Irish Fiddle Performance Model Informed with Bowing." In *Proceedings of the International Computer Music Conference (ICMC 2008)*, Sonic Arts Research Centre: Belfast, Northern Ireland.
Puiggròs, Montserrat, Emilia Gómez, Rafael Ramírez, Xavier Serra, and Roberto Bresin. 2006. "Automatic Characterization of Ornamentation from Bassoon Recordings for Expressive Synthesis." In *Proceedings of 9th International Conference on Music Perception and Cognition*, 22–26 August 2006, University of Bologna, Italy.
Ramírez, Rafael, and Amaury Hazan. 2006. "A Tool for Generating and Explaining Expressive Music Performances of Monophonic Jazz Melodies." *International Journal on Artificial Intelligence Tools* 15 (4): 673–691.
The Real Book Series, Milwaukee, WI: Hal Leonard. 2013. http://www.halleonard.com/aboutUs.jsp.

Serra, Xavier. 1997. "Musical Sound Modeling with Sinusoids Plus Noise." *Musical Signal Processing* 91–122.
Todd, Neil. 1989. "A Computational Model of Rubato." *Contemporary Music Review* 3 (1): 69–88.
Todd, Neil P. McAngus. 1992. "The Dynamics of Dynamics: A Model of Musical Expression." *The Journal of the Acoustical Society of America* 91 (6): 3540–3550.
Todd, Neil P. McAngus. 1995. "The Kinematics of Musical Expression." *The Journal of the Acoustical Society of America* 97 (3): 1940–1949.
Widmer, Gerhard. 2003. "Discovering Simple Rules in Complex Data: A Meta-Learning Algorithm and Some Surprising Musical Discoveries." *Artificial Intelligence* 146 (2): 129–148.
Widmer, Gerhard, and Asmir Tobudic. 2003. "Playing Mozart by Analogy: Learning Multi-Level Timing and Dynamics Strategies." *Journal of New Music Research* 32 (3): 259–268.
Woodrow, Herbert. 1951. "Time Perception." In *Handbook of Experimental Psychology*, edited by Stanley Smith Stevens, 1224–1236. New York: Wiley.
Zapata, José R., André Holzapfel, Matthew E. P. Davies, João Lobato Oliveira, and Fabien Gouyon. 2012. "Assigning a Confidence Threshold on Automatic Beat Annotation in Large Datasets." In *13th International Society for Music Information Retrieval Conference*, 8–12 October 2012, Porto, Portugal, 157–162.

Analysis of analysis: Using machine learning to evaluate the importance of music parameters for Schenkerian analysis

Phillip B. Kirlin and Jason Yust

While criteria for Schenkerian analysis have been much discussed, such discussions have generally not been informed by data. Kirlin [Kirlin, Phillip B., 2014 "A Probabilistic Model of Hierarchical Music Analysis." Ph.D. thesis, University of Massachusetts Amherst] has begun to fill this vacuum with a corpus of textbook Schenkerian analyses encoded using data structures suggested byYust [Yust, Jason, 2006 "Formal Models of Prolongation." Ph.D. thesis, University of Washington] and a machine learning algorithm based on this dataset that can produce analyses with a reasonable degree of accuracy. In this work, we examine what musical features (scale degree, harmony, metrical weight) are most significant in the performance of Kirlin's algorithm.

1. Introduction

Schenkerian analysis is widely understood as central to the theory of tonal music. Yet, many of the most prominent voices in the field emphasize its status as an expert practice rather than as a theory. Burstein (2011, 116), for instance, argues for preferring a Schenkerian analysis "not because it demonstrates features that are objectively or intersubjectively present in the passage, but rather because I believe it encourages a plausible yet stimulating and exciting way of perceiving and performing [the] passage." Rothstein (1990, 298) explains an approach to Schenker pedagogy as follows.

> Analysis should lead to better hearing, better performing, and better thinking about music, not just to "correct" answers. [...] I spend lots of class time—as much as possible—debating the merits of alternate readings: not primarily their conformance with the theory, though that is discussed where appropriate, but their relative plausibility as models of the composition being analyzed.

Schachter (1990) illustrates alternative readings of many works and asserts that a full musical context is essential to evaluating them. Paring the music down to just aspects of harmony and voice leading, like "the endless formulas in white notes that disfigure so many harmony texts," he claims, leaves the difference between competing interpretations undecidable. In publications such deliberation typically occurs at a high level. It rarely addresses the implicit principles used to deal with many details of the musical surface. As Agawu (2009, 116) says, "the journey from strict counterpoint to free composition makes an illicit or – better – mysterious leap as it approaches its destination."

As with any complex human activity, the techniques of artificial intelligence may greatly advance our understanding of how Schenkerian analysis is performed and what kinds of implicit cognitive abilities and priorities support it. The present work builds upon the research of (Kirlin 2014a; 2015) which models Schenkerian analysis using machine learning techniques. By probing Kirlin's algorithm we address a question of deep interest to Schenkerian analysts and pedagogues: what roles do different aspects of the music play in deliberating between possible analyses of the same musical passage? Because the activity of Schenkerian analysis involves such a vast amount of implicit musical knowledge, it is treacherous to litigate this question by intuition, without the aid of computational models and methods.

This article is organized as follows. The next section, numbered 2, explains the machine learning algorithm we used, which is essentially that of Kirlin (2014b). Section 3 describes a series of experiments to test which musical features the algorithm relies upon most heavily to produce accurate analyses. Section 4 describes the results of that experiment, and Section 5 provides further exploratory analysis of data produced by the experiment.

2. A machine learning algorithm for Schenkerian analysis

Schenkerian theory is grounded in the idea that a tonal composition is organized as a hierarchical collection of *prolongations*, where a prolongation is defined, for our purposes, as an instance where a motion from one musical event, *L*, to another non-adjacent event, *R*, is understood to control the passage between *L* and *R*, and the intermediate events it contains. A prolongation is represented in Schenkerian notation as a slur or beam.

Consider Figure 1, a descending melodic passage outlining a G major chord. Assuming this melody takes place over G-major harmony, this passage contains two passing tones (non-harmonic tones in a melodic line that linearly connect two consonant notes via stepwise motion), the second note C and the fourth note A. These tones smoothly guide the melody between the chord tones D, B, and G. In Schenkerian terminology, the C *prolongs* the motion from the D to the B, and the A similarly *prolongs* the motion from the B to the G.

The hierarchical aspect of Schenkerian analysis comes into play when we consider a prolongation that occurs over the entire five-note passage. The slurs from D to B and B to G identify the C and A as passing tones. Another slur from D to G, which contains the smaller slurs, shows that the entire motion outlines the tonic triad from D down to G. The placement of slurs may reflect the relatively higher stability of the endpoints (between chord tones over non-chord tones, and between more stable members of the triad, root and fifth, over the third), or they may reflect a way in which the local motions (passing-tone figures) group into the most coherent larger-scale motion (arpeggiation of a triad).

Figure 1. A melodic line illustrating prolongations.

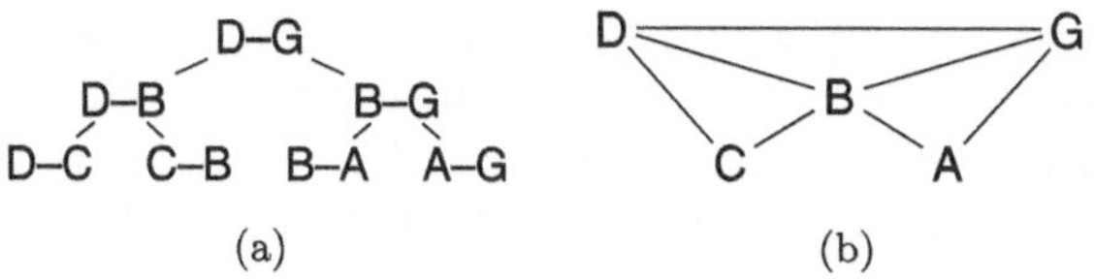

Figure 2. The prolongational hierarchy of a G-major chord with passing tones represented as two equivalent data structures. (a) A binary tree of melodic intervals. (b) A maximal outerplaner graph, or MOP.

This hierarchy can be represented visually by the tree in Figure 2(a): this diagram illustrates the hierarchy of melodic intervals present in the composition and the various prolongations identified above. An equivalent representation, known as a *maximal outerplanar graph*, or MOP (Yust 2006, 2009, 2015), is shown in Figure 2(b). Binary trees of intervals and MOPs are duals of each other in that they represent identical sets of information, though the MOP representation is more succinct.

From a mathematical perspective, a MOP is a complete triangulation of a polygon. Each triangle in a MOP represents a single melodic prolongation among the three notes of the music represented by the three endpoints of the triangle. Because MOPs are oriented temporally, with the notes shown in a MOP always ordered from left to right as they are in the musical score, we can unambiguously refer to the three endpoints of a triangle in a MOP as the left (L), middle (M), and right (R) endpoints. Each triangle in a MOP, therefore, represents a prolongation of the melodic interval from L to R by the intervals from L to M and M to R.

2.1. *A probabilistic interpretation of MOPs*

Our goal is to develop an algorithm with the ability to predict, given a musical composition, a "correct" Schenkerian analysis for that composition. We develop this algorithm using the following probabilistic perspective.

Assume that we are given a sequence of notes N that we wish to analyze, and that all possible Schenkerian analyses of N can be enumerated as $A_1, \ldots, A_m$, for some integer m. We desire the most probable analysis given the notes, which is $\arg\max_i P(A_i \mid N)$. Because a Schenkerian analysis A_i can be represented in MOP form by its collection of triangles, we will denote the set of triangles comprising analysis A_i by $T_{i,1}, \ldots, T_{i,p}$, for some integer p. We then define $P(A_i \mid N)$ as the joint probability $P(T_{i,1}, \ldots, T_{i,p})$. In other words, we define the probability of an analysis being correct for a given sequence of notes as the probability of observing a MOP containing the specific set of triangles derived from that analysis.

We will use supervised machine learning to estimate this probability from a corpus of Schenkerian analyses done by humans, which we interpret as ground truth. Specifically, we use the SCHENKER41 dataset, the largest known collection of machine-readable Schenkerian analyses in existence (Kirlin 2014a). This dataset contains 41 common-practice era musical excerpts and their corresponding Schenkerian analyses. All of the excerpts are either for a solo keyboard instrument (or arranged for such an instrument) or for voice with keyboard accompaniment. All are in major keys and do not modulate, though there are some tonicizations. The musical excerpts are encoded in the symbolic MusicXML format, whereas the corresponding analyses are encoded in a text-based format that captures the prolongational information in a Schenkerian analysis, while not assigning any musical interpretation to each prolongation. In other words, the prolongations are not labeled with terms such as "passing tone," "neighbor tone," or other descriptors, but are denoted only by their constituent notes. Each analysis also contains a Roman numeral harmonic labelling, determined either from the same source as the prolongations themselves, or from an expert music analyst.

The analyses in the dataset are derived from four sources: Forte and Gilbert's *Introduction to Schenkerian Analysis* and the accompanying solutions manual (1982a; 1982b) (10 excerpts), Cadwallader and Gagné's *Analysis of Tonal Music* (1998) (4 excerpts), Pankhurst's *Schenker-GUIDE* (2008) (24 excerpts), and teaching materials produced by an expert music analyst (3 excerpts).

Each excerpt in the dataset can be translated into a MOP representing the prolongations present in the main melody of the excerpt. However, it is difficult to estimate the full joint probability $P(T_{i,1}, \ldots, T_{i,p})$ directly from the resulting collection of MOPs owing to the curse of dimensionality: the large number of possible combinations of triangles is simply too large for any reasonably-sized dataset. Instead, we make the simplifying assumption that each triangle in a MOP is independent of all other triangles in a MOP, which implies $P(T_{i,1}, \ldots, T_{i,p}) = P(T_{i,1}) \cdots P(T_{i,p})$. This assumption reduces the full joint probability to a product of simpler, lower-dimensional probabilities, which are easier to learn. An experiment verifies that this assumption largely preserves relative probability scores between two MOPs, which is a sufficient condition for our purposes to proceed (Kirlin and Jensen 2011).

Finally, we define the probability of an individual triangle appearing in a MOP as the probability of a given melodic interval being elaborated by the specific choice of a certain child note. That is, we define $P(T_{i,j}) = P(M_{i,j} \mid L_{i,j}, R_{i,j})$, where $L_{i,j}$, $M_{i,j}$, and $R_{i,j}$ are the three endpoints (notes) of triangle $T_{i,j}$. In summary, we have defined the most likely analysis A_i for a given sequence of notes N as

$$\begin{aligned}\arg\max_i P(A_i \mid N) &= \arg\max_i P(T_{i,1}, \ldots, T_{i,p}) \\ &= \arg\max_i \prod_{j=1}^{p} P(T_{i,j}) \\ &= \arg\max_i \prod_{j=1}^{p} P(M_{i,j} \mid L_{i,j}, R_{i,j}). \end{aligned} \tag{1}$$

We now return to the question of using the SCHENKER41 corpus to compute the probability $P(M_{i,j} \mid L_{i,j}, R_{i,j})$, which requires us to define exactly what type of musical information underlies the variables $L_{i,j}$, $M_{i,j}$, and $R_{i,j}$. It is typical to describe these variables by a collection of *features*: specific measurable properties of the source music whose values are determined from the notes of the music corresponding to those variables. We use a set of 18 features that provide basic melodic, harmonic, metrical, and temporal information about the music; these are described more specifically in Section 3.

With a set of features in hand, a straightforward approach to determining the probability above involves counting the frequency of every type of triangle within the corpus, where each type of triangle is determined from a unique combination of the 18 features. The curse of dimensionality thwarts us again: there are too many possible types of triangle to expect to get a reasonable frequency count from any reasonably-sized corpus of analyses. Instead, we use *random forests* (Breiman 2001), a machine learning ensemble method particularly suited for high-dimensional feature spaces, to learn the conditional probability $P(M_{i,j} \mid L_{i,j}, R_{i,j})$. Random forests, which operate by constructing a large collection of decision trees, are typically used for classification tasks by outputting the most frequent class (the mode) predicted by the collection of trees. However, a probability distribution can be obtained instead by counting the frequencies of the predicted classes and normalizing them (Provost and Domingos 2003).

Our set of 18 features includes a subset of six features that depend on the middle note $M_{i,j}$. It is unlikely that a single random forest could sufficiently learn to predict all six features simultaneously. Therefore, we factor our conditional probability into the product of six different probabilities, each assigned to predict one of the six features of the middle note. If we denote these six features of the middle note as M_1 through M_6, this factorization becomes (dropping the i, j subscripts for brevity):

$$\begin{aligned} P(M \mid L,R) &= P(M_1,M_2,\ldots,M_6 \mid L,R) \\ &= P(M_1 \mid L,R) \cdot P(M_2,M_3,\ldots,M_6 \mid M_1,L,R) \\ &= P(M_1 \mid L,R) \cdot P(M_2 \mid M_1,L,R) \cdot P(M_3,M_4,\ldots,M_6 \mid M_1,M_2,L,R) \\ &= P(M_1 \mid L,R) \cdot P(M_2 \mid M_1,L,R) \cdots P(M_6 \mid M_1,\ldots,M_5,L,R). \end{aligned}$$

In other words, we construct six random forests and use each one to construct a conditional probability distribution over a single feature of the middle note. Multiplying these distributions together gives us a distribution over all six features of the middle note, and therefore an estimate of the probability of seeing any particular triangle in a MOP.

Now that we have an appropriate estimate of this probability $P(M_{i,j} \mid L_{i,j}, R_{i,j})$, we can calculate the probability of an entire MOP analysis according to equation (1). Computationally, it is inefficient to enumerate all possible Schenkerian analyses for a given sequence of notes; the size of this set grows exponentially with the length of the sequence. Instead, by combining the probabilistic interpretation of MOPs along with the equivalence between MOPs and binary trees, we can view each prolongation within a MOP as a production in a probabilistic context-free grammar. Under this interpretation, it is straightforward to use standard parsing techniques (Jiménez and Marzal 2000; Jurafsky and Martin 2009) to develop an $O(n^3)$ algorithm, known as PARSEMOP-C, that determines the most probable analysis for a given sequence of notes. Additionally, the grammar formalism allows us to restrict the predicted analyses to those that contain a valid *Urlinie* through specific sets of production rules.

PARSEMOP-C accepts as input a sequence of notes to analyze, along with information about the harmonic context of those notes and what category of *Urlinie* should be located (e.g., $\hat{5}$-line or $\hat{3}$-line). The algorithm begins by computing the probabilities of analyses of all three-note excerpts of the input music. Because a three-note excerpt forms exactly one triangle in a MOP, these probabilities come directly from the output of the random forests. The most probable analyses for all four-note excerpts are obtained by combining the pre-computed results for three-note excerpts. This continues with n-note excerpts being analyzed by combining probabilities of $(n-1)$-note excerpts, until the entire piece is analyzed. The most probable analysis is output as a MOP, though there is an additional algorithm available that can display the analysis in more traditional Schenkerian notation.

We can compare the predicted output analyses of PARSEMOP-C against the ground-truth analyses from the SCHENKER41 corpus to determine the performance of the algorithm, which we measure via *edge accuracy*, which is the percentage of edges in an algorithmically-produced MOP that correspond to edges in the ground-truth MOP. Additional details on the probabilistic interpretation of MOPs, PARSEMOP-C, and its evaluation are available in Kirlin and Jensen (2015) and Kirlin (2014b).

PARSEMOP-C is the first completely data-driven Schenkerian analysis algorithm, in that it learns all the "rules" of Schenkerian analysis, including their forms and where the apply them, completely from a corpus of data. Very early studies in computional approaches to Schenkerian analysis encountered difficulties due to the issues inherent in handling the seemingly conflicting guidelines of the analysis process (Kassler 1975, 1987; Frankel, Rosenschein, and Smoliar 1978). Lerdahl and Jackendoff's *A Generative Theory of Tonal Music* (1983) presented two different formal systems for music that, like Schenkerian analysis, can be used to view a piece of music as a hierarchy of musical objects. The authors, however, stated that their formalizations were not designed to replicate the ideas of Schenkerian analysis, but were rather a new investigation in musical cognition. Though Lerdahl and Jackendoff's systems lack the details necessary to develop a "computable procedure for determining musical analyses," a number of further endeavors attempted to fill in the gaps. The most successful of these is the work done by Hamanaka et al., which has resulted in a collection of computational systems that attempt to replicate parts of Lerdahl and Jackendoff's theories. Their early systems (Hamanaka, Hirata, and Tojo 2005, 2006, 2007) required user-supplied parameters to identify the optimal music analysis, while later systems (Hamanaka and Tojo 2009; Miura et al. 2009; Kanamori, Hamanaka, and Hoshino 2014) have focused on using machine learning to automate the search for the appropriate parameters. Numerous other approaches (Mavromatis and Brown 2004; Gilbert and Conklin 2007; Marsden 2005; Marsden and Wiggins 2008; Marsden 2010) have used a context-free grammar formalization similar to the one presented above, but with a hand-created rule set rather than rules learned directly from the ground-truth analyses.

3. Musical features

To understand how individual musical features might figure into the accuracy of these algorithmic analyses, consider the task of analyzing the "music" in Figure 3. This is the intervallic pattern of the melody of a simple four-measure phrase of real music. There is no information about the rhythm or harmony of the original music. We do not know what the key is or which note is the tonic. Registral information is also removed by octave-reducing melodic intervals and inverting larger leaps. How likely would it be that one could reproduce a textbook analysis of the melody relying on this intervallic pattern only?

Or what if one were given only the rhythm and meter of a melody, as in Figure 4? How accurately could we expect to predict a Schenkerian analysis of the passage?

In the experiment described below, we train the machine learning algorithm on Schenkerian analyses with differing amounts of information about the music, and therefore discover how important different aspects of the music (melodic, harmonic, metric) are to accurately reproducing Schenkerian analyses.

The data for Figures 3 and 4 comes from Schubert's Impromptu Op. 142, No. 2, which is analyzed by Cadwallader and Gagné (1998) as shown in Figure 5. Their analysis has enough detail to be translated into the explicit and near-complete MOP shown in Figure 6(**a**). The MOP is a simplification, reflecting just the basic melodic hierarchy implied by Cadwallader and Gagné's analysis. There are three places where their analysis is ambiguous: in measures 1–3, they show a double neighbor figure A♭-G-B♭-A♭, where a slur could be added from A to B♭ or from G to A♭, but neither is clearly implied by the analysis. In measures 5–8, they show C-B♭-A♭-C in stemmed notes. A slur from C to A♭ might be inferred but is not actually shown, so it is not included in the coded analysis. Finally, this example reflects a common problem in that Cadwallader and Gagné give the status of $\hat{3}$ to two different Cs. A fully explicit analysis would choose between

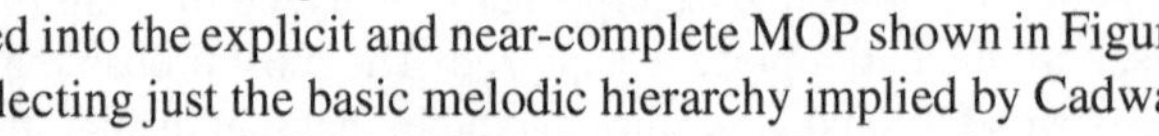

Figure 3. A melody with general interval information only.

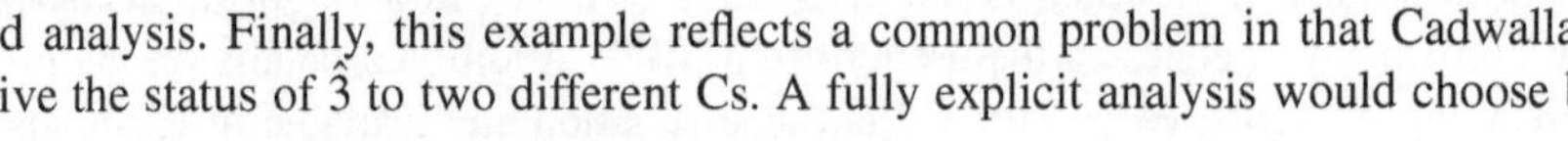

Figure 4. The meter and rhythm of a melody.

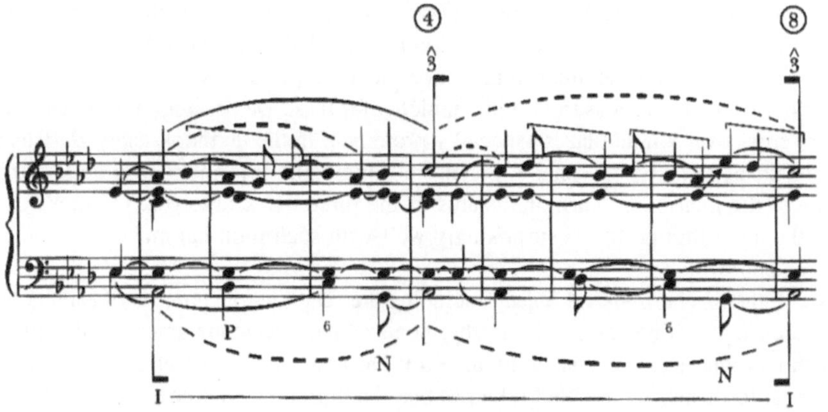

Figure 5. Cadwallader and Gagné (1998)'s analysis of Schubert's Impromptu Op. 142, No. 2.

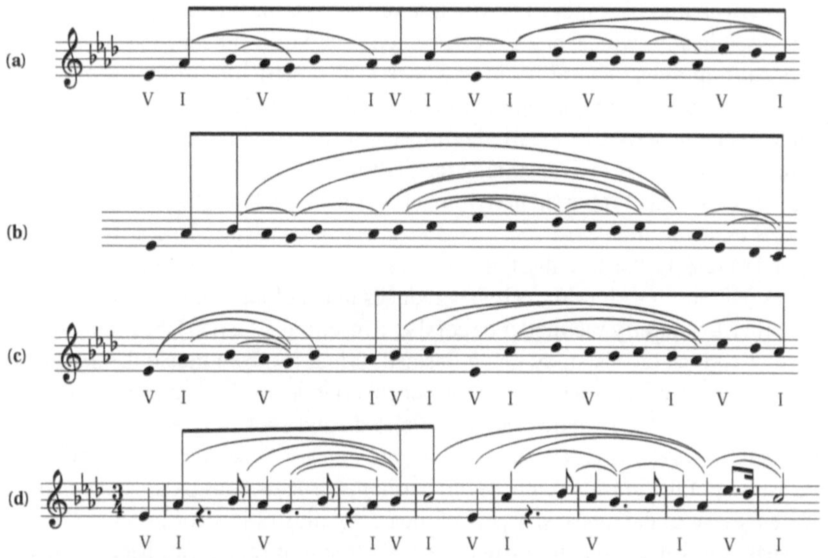

Figure 6. (**a**) An encoded version of the analysis in Figure 5; (**b**) an analysis of just the intervallic pattern of the impromptu, as shown in Figure 3, by a computer trained on just the intervallic patterns of textbook analyses; (**c**) computer analysis of the melodic (interval and scale degree) and harmonic information; and (**d**) a computer analysis of the melodic, harmonic, and metrical data of the impromptu.

these as the true initiation of the fundamental line. The part of the analysis shown with beams is the "background" of the analysis; the ParseMop-C algorithm is constrained to find such a $\hat{1}$-$\hat{2}$-$\hat{3}$ initial ascent as the background of the phrase (but not to find these notes in any specific locations).

Figures 6(**b**)–6(**d**) show three analyses by the algorithm trained on limited amounts of musical information, which can be compared to the "ground truth" in Figure 6(**a**). The first, Figure 6(**b**), shows the solution arrived at by the algorithm trained on just the generic melodic intervals of the corpus of textbook analyses. Since this version of the algorithm has no awareness of keys, accidentals, harmony, register, or rhythm, the music is shown with no key signature, clef, or durational values, and with octave-reduced generic intervals (as in Figure 3). Note that adjacent notes are implicitly connected, so a slur between the first two notes would be redundant. The algorithm accurately identifies some local passing motions in the music given just this basic generic intervallic information. Evaluated by the proportion of shared slurs between the computer analysis and the textbook one, however, the computer does not do especially well on this example, mostly because it buries the C of measure 4 in the middle of a B♭-C-D♭ passing motion, and thus most of its slurs cross over a note that is especially prominent in the textbook analysis. This passing motion, which is perfectly reasonable given only intervallic information, is quite implausible when we see the metric and harmonic status of the three notes involved. On other examples, however, the algorithm performs surprisingly well with such minimal musical information, as we shall see below.

Given a slightly richer musical object, including the scale-degree number of each note (i.e., it has the reference point of a key), and some harmonic context (Roman numeral without inversion) the algorithm produces the analysis of Figure 6(**c**). The new information allows it to avoid certain blatant errors: for instance, now that it knows that the third note (B♭) occurs over a tonic chord, not a V, it avoids assigning it a major structural role. However, the algorithm makes another decision that turns out to be a mistake, shifting the first note of the structure ahead to note 7. It apparently identifies an arpeggiation of V as a likely introductory approach to the first structural note, but that turns out to be implausible because of the rhythm. When it has additional metrical information (specifically, which notes are accented relative to others), the algorithm corrects this mistake, shifting the structural beginning back to the metrically strong note 2, as shown in Figure 6(**d**). This result is then quite similar to Cadwallader and Gagné's analysis.

This single example shows how different aspects of the music – melodic pattern, harmony, rhythm – play different roles in determining the plausibility of a Schenkerian analysis. The principal goal of the experiment reported here is to answer the broad question of *what features of the music are most essential to accurately reproducing human analyses.* To clarify, this experiment is not designed to select a "best" subset of variables that will be used in future studies; using random forests as our learning algorithm already reduces much of the need to find an ideal subset of variables in order to simplify the model or combat overfitting (Hastie, Tibshirani, and Friedman 2009). The motivations of this study come, instead, primarily from music theory. For instance, theorists often emphasize the linear aspect of Schenkerian analysis, to distinguish it from other analytical methods that focus more heavily on chords and root progressions. Without a data-driven approach, however, it is not really clear how much a Schenkerian analysis really relies on melodic aspects of the music as opposed to harmonic progressions. Also, Schenker pedagogy typically de-emphasizes the metrical aspect of music, by illustrating, for instance, how brief and metrically weak notes may sometimes play central structural roles. Yet, because analysts may rely upon metrical information unconsciously, in making less prominent analytical decisions, such arguments may be misleading. Finally, Schenkerian analysis is often tacitly understood to be a recursive procedure, meaning that the same criteria apply to very local analytical decisions as those at higher levels, between non-adjacent notes and measure-to-measure across a phrase. Although the examples in the SCHENKER41 corpus are primarily short – usually just a single phrase or pair of phrases – we were able to test a limited version of this hypothesis from the note-to-note to the measure-to-measure level, by including a set of temporal features, showing that even this limited version of the recursion hypothesis is incorrect.

The main results (reported in Section 4) are the outcomes of two feature selection processes. The first tests the relative importance of broadly-categorized harmonic, melodic, metric, and temporal features. The second is an ordering of the more specific set of 18 features according to how much the performance of the machine learning algorithm depends upon them. This second process generated a substantial amount of data that also permitted us to draw some limited conclusions via exploratory analysis of how features interact. This is reported in Section 5.

The features used in this work were originally specified by Kirlin (2014b) because they covered a range of types of musical information, melodic, harmonic, rhythmic, and temporal, were easily determinable by computer from the dataset, had a small number of possible values for each feature, and most importantly, seemed likely to fulfill a critical role in the Schenkerian analytical process. The features also proved well-suited to test the kinds of common assumptions about Schenkerian analysis described above.

We define 18 features in all that are available to the algorithm in the full model. The following six of these are features of the middle note.

- **SD-M** The scale degree of the note (represented as an integer from 1 through 7, qualified as raised or lowered for altered scale degrees).
- **RN-M** The harmony present in the music at the time of onset of the center note (represented as a Roman numeral from I through VII or "cadential six-four"). For applied chords (tonicizations), labels correspond to the key of the tonicization.
- **HC-M** The category of harmony present in the music at the time of the center note, represented as a selection from the set *tonic* (any I chord), *dominant* (any V or VII chord), *predominant* (II, II^6, or IV), *applied dominant*, or *VI chord*. (Our dataset did not have any III chords.)
- **CT-M** Whether the note is a chord tone in the harmony present at the time (represented as a selection from the set "basic chord member" (root, third, or fifth), "seventh of the chord," or "not in the chord").
- **Met-LMR*** The metrical strength of the middle note's position as compared to the metrical strength of note *L*, and to the metrical strength of note *R* (represented as a selection from the set "weaker," "same," or "stronger").
- **Int-LMR*** The melodic intervals from *L* to *M* and from *M* to *R*, generic (scale-step values) and octave generalized (ranging from a unison to a seventh).

Two features of the middle note are different than the others, in that they are influenced by the left and right notes. These are therefore distinguished as "LMR" features, as opposed to simple "M" features.

We also used the following 12 features for the left and right notes, *L* and *R*.

- **SD-LR*** The scale degree (1–7, qualified as in SD-M) of the notes *L* and *R*.
- **Int-LR** The melodic interval from *L* to *R*, with intervening octaves removed.
- **IntI-LR** The melodic interval from *L* to *R*, with intervening octaves removed and intervals larger than a fourth inverted.
- **IntD-LR** The direction of the melodic interval from *L* to *R*; i.e., up or down.
- **RN-LR*** The harmony present in the music at the time of *L* or *R*, represented as a Roman numeral I through VII, or "cadential six-four."
- **HC-LR*** The category of harmony present in the music at the time of *L* or *R*, represented as a selection from the set tonic, dominant, predominant, applied dominant, or VI chord.
- **CT-LR*** Status of *L* or *R* as a chord tone in the harmony present at the time (see the description of CT-M above).
- **MetN-LR*** A number indicating the beat strength of the metrical position of *L* or *R*. The downbeat of a measure is 0. For duple or quadruple meters, the halfway point of the measure is 1; for triple meters, beats two and three are 1. This pattern continues with strength levels of 2, 3, and so on.
- **MetO-LR*** A number indicating the beat strength of the metrical position of *L* or *R*. The downbeat of a measure is 0. For duple or quadruple meters, the halfway point of the measure is 1; for triple meters, beats two and three are 1. This pattern continues with strength levels of 2, 3, and so on. The difference between MetN-LR and MetO-LR is that the former is represented in trees of the random forest as a *numeric* variable, and the latter as an *ordinal* variable.

The algorithm that creates the forests treats these types of variables differently: numeric variables are compared using greater-than or less-than tests, while ordinal variables are treated as categorical, and are compared using equality tests.

- **Lev1-LR** Whether *L*, *M*, and *R* are consecutive notes in the music, represented as a true/false value (this can be determined strictly from examining the positions of *L* and *R*, and so is not included in the list of middle-note features).
- **Lev2-LR** Whether *L* and *R* are in the same measure in the music, represented as a true/false value.
- **Lev3-LR** Whether *L* and *R* are in consecutive measures in the music, represented as a true/false value.

Note: Features marked with an asterisk (*) in the lists above are, from a machine learning standpoint, two different features: they are separate and distinct inputs available to the random forests algorithm; for instance, there is an SD-L feature and an SD-R feature that the learning algorithm treats independently. However, in the feature selection experiments that follow, these "paired features" are always treated as an indivisible unit in the set of available features– they are always added or removed together.

There are many well-established feature selection techniques that could potentially help determine the relative importance of the features in producing output analyses that match the ground-truth analyses. Many are inappropriate in our situation, however, because they are designed specifically for classification tasks, and there is no notion of an output class in our system. A further complication is that any change in the input features directly affects the output of the random forests, which is a probability distribution for which we do not have ground-truth data. Certainly, improving the estimate of the probability distribution will likely improve the quality of the most-probable music analysis predicted by ParseMop-C, but it is unclear to what degree this is true, as the two algorithms are computationally separate. Therefore, we turned to two standard subset selection methods along with examining correlations between features.

In the two experiments that follow, we evaluated the quality of a specific subset of features by running the ParseMop-C algorithm with leave-one-out cross-validation to compute new MOP analyses for all the pieces in the Schenker41 corpus longer than four measures, then calculated the overall edge accuracy for all the MOPs combined. For our first experiment, we divided the features into four broad categories, melodic, harmonic, rhythmic, and temporal, and evaluated the quality of all exhaustive subsets of these categories of features. In our second experiment, we used *backward selection* to cycle through each feature from the training data and in order to find the feature that, when omitted, decreased the overall accuracy the least. This feature was then permanently removed and another cycle was performed, until only one feature remained. For the entire set of 18 features we then had 18 trials, numbered 0–17, where trial 0 included all 18 features in the full model, and trial 17 (trivially) included only the last remaining feature.

4. Results

4.1. *Experiment 1*

We first divided the features into four categories as follows.

- Melodic: SD-M, SD-LR, Int-LMR, Int-LR, IntI-LR, IntD-LR
- Harmonic: RN-M, RN-LR, HC-M, HC-LR, CT-M, CT-LR
- Metrical: Met-LMR, MetN-LR, MetO-LR
- Temporal: Lev1-LR, Lev2-LR, Lev3-LR

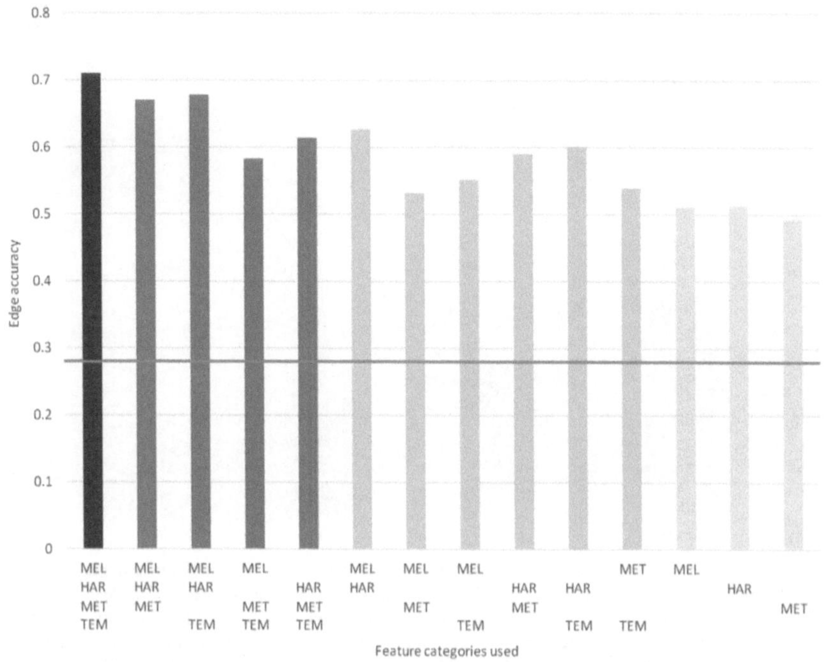

Figure 7. Edge accuracy as determined from different subsets of feature categories. There is no trial for the temporal features by themselves, as there are no middle-note features in that category, and at least one such feature is required. The horizontal line at 28% indicates the baseline edge accuracy that could be obtained by an algorithm that operated randomly.

We ran 14 trials to evaluate the quality of non-zero sized subsets of feature categories, from all four categories (including all 18 features) down to single categories. Each trial produced an overall edge accuracy giving us information on the utility of the categories of features, as detailed in Figure 7. The random forests use the left–right features as "input" and the middle-note features as "output," and therefore trials are impossible to run if no middle-note features are included. Thus, there is no "temporal-only" trial.

Figure 7 shows a baseline edge accuracy of roughly 28%. This number is calculated by averaging the edge accuracies of all possible MOP analyses of all the musical excerpts in question. In other words, a hypothetical version of ParseMop that analyzed music by selecting a MOP analysis uniformly at random from all possibilities would average 28% edge accuracy.

The data from this experiment allow us to test three broad music-theoretic questions described in Section 3: first, what is the relative importance of melodic and harmonic information? Second, is metrical information necessary for producing accurate analyses? Third, does temporal information influence analyses – that is, is it necessary to use harmonic, melodic, and metrical information differently depending on whether one is making decisions at note-to-note or measure-to-measure levels? All of these questions essentially presume that each feature set has a relatively consistent influence over the performance between trials, regardless of the other feature sets present. Figure 8 therefore shows the change of performance between trials that differed only in the presence or absence of one feature set, melodic, harmonic, metrical, or temporal. These appear to reflect relatively normal distributions, so we further assessed the contribution of each feature set performing paired t-tests comparing data with and without the given feature, and found the inclusion of harmonic, melodic, and temporal features made a significant difference at $p < .001$ and metrical features at $p < .01$.

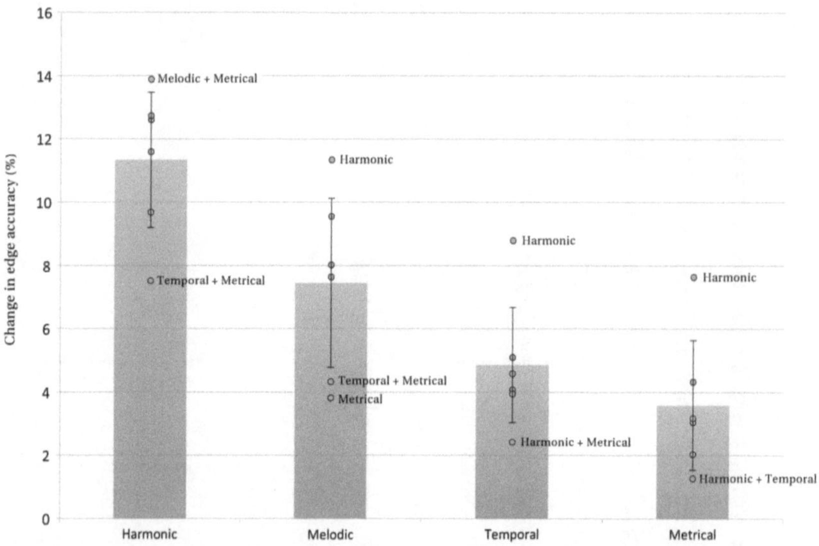

Figure 8. Change of performance for adding each feature set. Each data point is the difference in performance between trials that include the feature set listed on the *x*-axis and trials with those features removed. Averages and standard deviations are shown for the addition of each group of features. Data points falling outside of a single standard deviation of the average are labeled with the other feature sets present on the trials compared. For instance, the largest difference observed is between the trial with Melodic + Metrical features, and the one with Melodic + Metrical + Harmonic features.

From Experiment 1 we can draw the following conclusions.

- Harmonic features are the most essential to producing an accurate analysis. In particular, they outweigh melodic features, although these also are very important. This is shown in the large average influence over performance exhibited by these feature groups in Figure 8.
- Temporal features have a reliably positive influence on performance, around 5% on average in Figure 8, meaning that the rules of Schenkerian analysis do differ from level to level. In particular, this influence seems to be present regardless of what other feature sets are present, so we can infer that harmonic, melodic, and metrical information all need to be used differently in some way between note-to-note, within-measure, and/or measure-to-measure levels.
- Metrical features also show a reliable influence over performance in Figure 8, although it is relatively small (around 3–4%).

The extreme values in Figure 8 are also suggestive. All feature sets have well above-average influence when just the harmonic features are present, which suggests that harmonic features are most useful in combination with at least one other feature type. At the same time, three of the unusually low values point to the fact that one trial, Harmonic + Temporal + Metrical, is substantially lower than what would be predicted by a simple additive model of how the features interact. In other words, there appears to be an especially sizeable overlap in analytically usable information for this particular combination of features.

4.2. *Experiment 2*

The second experiment compared the expendability of the 18 individual features using the backward selection procedure. Figure 9 lists the features dropped on each trial. At certain points in the process there are larger changes in baseline edge accuracy, most notably on trials 14, 16, and 17.

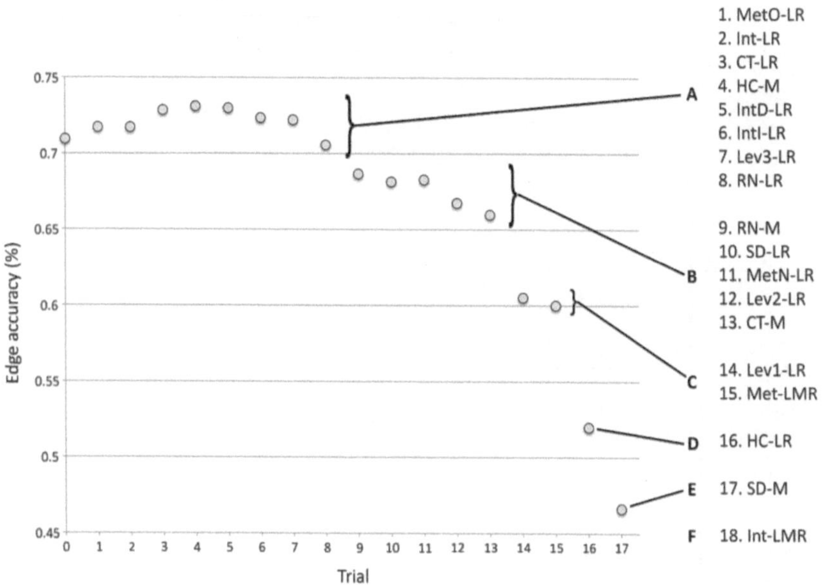

Figure 9. Performance of the algorithm on trials 0–17 and the feature removed prior to each trial. (Trial 0 includes all features.) N.B. There is no trial 18, but Int-LMR is listed as number 18 to show that it is the last feature remaining in the model.

This suggests that between these points are more significant differences in feature importance, and the features can be sorted into groups between these critical points.

As a check on the robustness of the ordering produced by the backward selection we also averaged, across all trials including the given feature, the decrease in performance observed after removing that feature. These data are shown in Figure 10. The orderings are consistent between the groupings indicated on Figure 9, but not within the larger groups A and B.[1] We also performed paired t-tests on the data comparing trials that differed only on the inclusion of the given feature. The results are shown in Figure 10. Note that n decreases as one goes higher on the list in Figure 9, and is very small for many of the features in group A. Nonetheless, the statistical tests show reliable influence over performance for all of the features in groups B–F, and even for some features in group A.

One striking aspect of the result is that, if we sort the features by type – melodic, harmonic, metrical, and temporal – the five features in groups C–F represent all four types, with a duplication for melodic features (Int-LMR and SD-M). In addition, the five features in group B also represent all types, with an additional harmonic feature (CT-M and RN-M). Since the ordering of groups C–F appears robust, while that within group B is not, we can broadly characterize the results as follows.

- The result of Experiment 2 is consistent with the finding from Experiment 1 that melodic, harmonic, metrical, and temporal information all contribute to the analytical process. It also indicates, further, that the algorithm performs best with a mix of all these feature types.
- It is also consistent with the finding from Experiment 1 that melodic and harmonic features are the most important (groups D–F), metrical and temporal somewhat less so (group C).

[1] These two measures are confounded: for the features that remain in the model for longer, there are more trials to average over, and the drop in performance predictably gets higher as the number of features in the full model gets smaller. The similarity of the two rankings can therefore be partly featured to this confound, but not entirely so. The average rise of feature importance from trial to trial is just 0.12%, and this is mostly due to the last four trials. On trials 0–13 the average change is 0.02%.

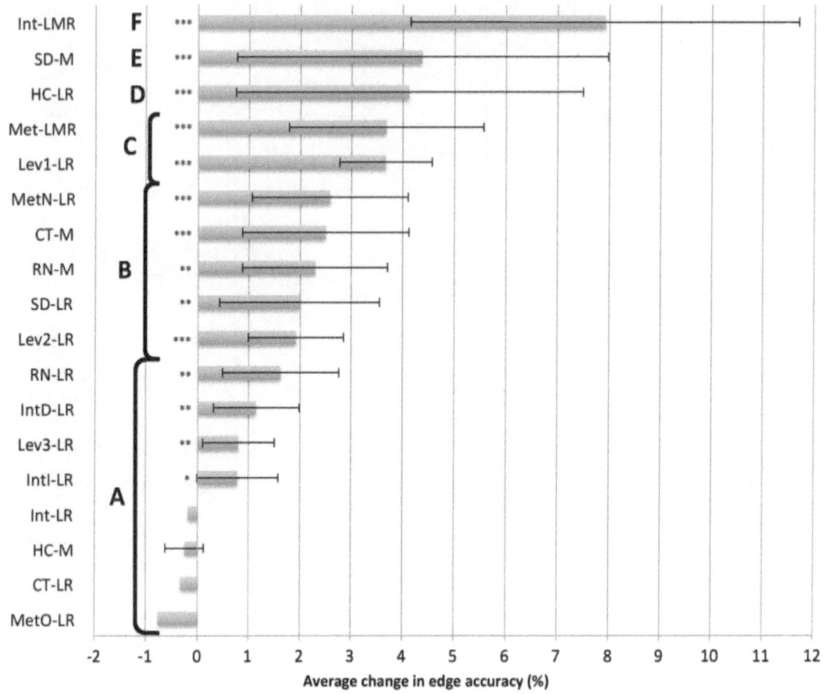

Figure 10. Average feature importance and standard deviation for all features over the course of the experiment. Asterisks show the results of paired *t*-tests between trials including and excluding the given feature. *** $= p < .001$, ** $= p < .01$, * $= p < .05$.

- For each of the feature types, except melodic, the majority of the necessary information is captured by a single feature. Yet, in all cases, some additional valuable information is provided by one or two other features (group B). In the case of melodic features, these can be subdivided into purely intervallic information (Int-LMR) and tonally anchored information (SD-M).

While the results of the two experiments are largely consistent, they give divergent indications on the relative significance of melodic and harmonic features. Experiment 1 indicated that harmonic features make a larger contribution to the analytical process than melodic ones, while in experiment 2, the highest-ranking features are melodic ones (Int-LMR and SD-M). Together, these two results indicate that the important melodic information is consolidated in these two specific features, which provide a melodic pattern (Int-LMR) and orient it to a tonal center (SD-M), whereas harmonic information is distributed among several different kinds of features: distinct RN/HC versus CT features, the former dividing up between M and LR types.

Other than SD-M and Int-LMR, the only other notable melodic feature is SD-LR (group B). For the most part, the scale degree of the left and right notes is predictable from the scale degree of the middle note and the intervals to and from the middle note. While SD-LR provides an additional distinction between chromatically altered scale degrees and diatonic ones, the significance of this feature probably is more due to the potential use of melodic information as an LR feature – i.e. on the left-hand side of the conditional probabilities. This is further discussed in Section 5.

Three of the harmony features drop out in group A. The HC-M and RN-M features, and HC-LR and RN-LR, are highly redundant (see Section 4.4 below), so it is unsurprising that one of each of these drops out in group A. The early exclusion of the CT-LR feature is also unsurprising, because non-chord-tones should be rare as LR features in the corpus. The other three harmony features all seem to be important: HC-LR, CT-M, and RN-M. HC-LR is the last LR feature (group D), and from trial 11 onward the only other LR features are temporal ones. This indicates that harmony provides the most useful context for predicting features of the middle note. CT-M is also a feature that would be expected to be important, especially at the note-to-note level, since non-chord-tones are usually afforded a low structural status. Some additional data analysis reported in Section 5 suggests that on trial 13, after the removal of CT-M, a group of other features (SD-M, HC-LR, Int-LMR, and MetN-LR) combine to predict the chord-tone status of the middle note effectively.

As in Experiment 1, the metrical and temporal features played a lesser but nonetheless significant role. Of the temporal features, both Lev1-LR and Lev2-LR were valuable, showing that it is mainly the note-to-note, within-measure, and between-measure levels that behave differently. The difference between consecutive measures and non-consecutive measures (indexed by Lev3-LR) was less useful. However, this may simply reflect the fact that the musical examples were relatively short, and important decisions at these higher levels were constrained by the given fundamental line.

Between the metrical features, it is unsurprising that Met-LMR is especially useful, since it incorporates relative information about the left and right notes as they relate to the middle note. However, MetN-LR, which provides absolute information about the metrical context, is also of some value.

4.3. *Experiment 3*

To complement the information about the relative expendability of features provided by the backward selection process in Experiment 2, we also performed a more limited forward selection process to assess the independent usefulness of smaller sets of features. Because M features require LR features, and vice versa, we paired all similar M and LR features, and also included the LMR features by themselves. The results are shown in Figure 11.

These results corroborated the usefulness of harmony, observed in Experiment 1 but less evident in Experiment 2. Melodic features alone give lower accuracy than either harmonic (not including CT) or metrical. This suggests that the high analytical value of melodic information is dependent upon some baseline harmonic and metrical information (in the form of HC-LR and Met-LMR), while harmony and meter are somewhat more independent. The especially low performance of the CT features can be attributed to the uselessness of CT-LR (given that the great majority of triangles in the corpus will have chord tones on the left and right), also observed in Experiment 2.

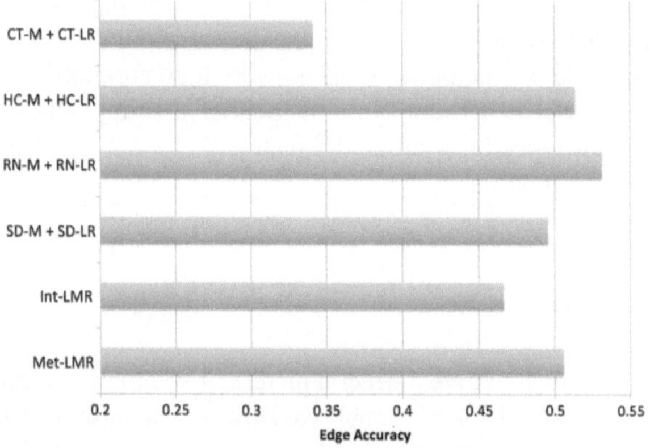

Figure 11. The edge accuracy of four models containing only a single LMR feature, or a single M and LR feature.

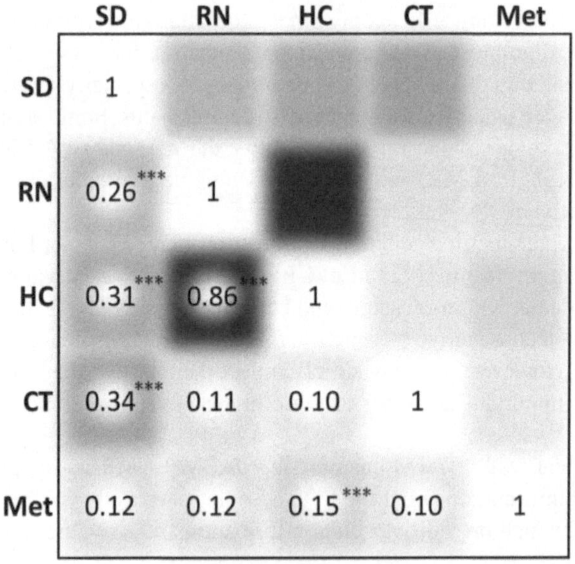

Figure 12. Cramer's *V* for five musical features (SD = scale degree, HC = harmony class, RN = Roman numeral, CT = chord tone status, and MS = metric strength). The numerical values are given below the diagonal and also visually illustrated by the relative darkness of squares in the grid. *** $= p < .001$.

4.4. *Correlations among features*

To aid in understanding how features may be interacting in the two main experiments, we also calculated correlations between features that existed in the musical excerpts themselves, without regard to the analyses (Figure 12). Having two features highly correlated with one another included in the model can be expected to reduce the overall influence of both features. Thus, knowing which features are correlated in the music can help to separate the influence of these correlations from redundancies that are specific to the analytical process.

The five features included are all those that can apply to a single note: distinctions having to do with relationships between notes are not included. There is one melodic feature (scale degree), three harmonic features (harmonic class, Roman numeral, and chord tone), and one metrical feature (metric strength). A number of correlations are significant at $p < .001$ (and no others were significant even at $p < .05$). Metric strength has no significant correlations except for a small one with harmony class. On the other hand, melodic information is strongly correlated with all of the harmonic features. This is unsurprising for RN and HC, but the high value for CT is more so. This high correlation is due primarily to the strong tendency of the "7th of chord" category to be represented by $\hat{4}$. Among the harmonic features, the very high correlation between HC and RN is unsurprising. These features differ mainly on the treatment of applied chords, which are rare, and otherwise on the greater specificity of the RN feature. CT is not significantly correlated with the other two harmonic features.

5. Interactions between features

The results in the previous section give a general picture of what kind of musical information is most important in producing a Schenkerian analysis. However, we also found that the importance of a given feature may be dependent upon what other features are present in the model. In some cases, the reasons for this are obvious. For instance, we found in Section 4.4 that scale degree and harmony class are strongly correlated in the music. Therefore, the presence of, e.g., SD-M should reduce the importance of HC-M (and SD-LR should reduce the importance of HC-LR, etc.). Where this reflects features of the music, it has nothing to do with Schenkerian analytical practice *per se*. However, similar kinds of redundancies might derive from the nature of the analytical process as well. And we also find that certain features *increase* in importance when another feature is included, which reflects an interdependence of features in the analytical process.

The backward selection process, because it required removing every remaining feature on every trial, produced data that made it possible to address this question of how features interact. The nature of this data analysis is necessarily exploratory. Consider, for instance, Figure 13, which tracks the effect of removing the scale degree feature over the course of Experiment 2. Were the relationship of this to other features equivocal, we might expect the values to stay roughly the same or to increase steadily from one trial to the next. The chart for SD-M is much more jagged, suggesting that it has non-trivial relationships with other features. For instance, when SD-LR is removed, SD-M suddenly spikes in importance. This suggests that the preceding low value probably reflected not a lack of significance for scale degree information in general, but that there is considerable overlap between the two scale degree features, making one of them – but not both – expendable. This makes sense, because given that the algorithm has intervallic information about the melody (in the form of Int-LMR), either scale degree feature can anchor that intervallic information to a tonal center.

The potentially most meaningful data in these graphs are the places like this where there are large changes in the importance of some feature from one trial to the next. We can infer that these large changes result from some kind of interaction between the feature in question and the one removed from one trial to the next. These may be of the form of redundancies or interdependencies. Such redundancies or interdependencies might involve more than one feature. For example, if one knows the scale degree of the middle note (SD-M) then one can infer the scale degree of the left and right notes (SD-LR) by knowing the intervals between the middle and the left and right notes (Int-LMR). Therefore, these three features are redundant as a group, but no pair of them is redundant.

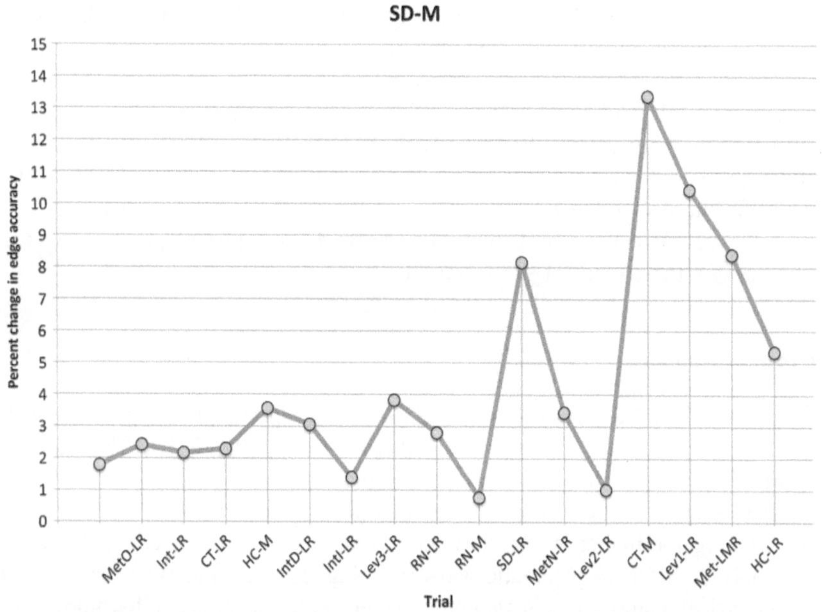

Figure 13. The effect of removing the SD-M feature on trials 0–16. The labels on the horizontal axis show the feature removed prior to the given trial. For instance, the peak at the trial for SD-LR means that the effect of removing SD-M changed to about 8% *after* the removal of SD-LR.

This redundancy is reflected by spikes in the importance of SD-M in Figure 13. The drops in importance reflect interdependencies between features. The largest drop in Figure 13 occurs when MetN-LR is removed. This indicates that SD-M is more useful when MetN-LR is included in the model. MetN-LR provides absolute metric orientation to complement the relative information of Met-LMR, so we can deduce that the likelihood of a given scale degree in the middle position changes depending on the metric context. We also see fairly large drops in the importance of SD-M for the temporal features Lev1-LR and Lev2-LR, which similarly indicates that the likelihood of certain scale-degree progressions depends on temporal context.

To search the data systematically for the most prominent such interactions between parameters, we tallied every change of feature importance from one trial to the next and ordered them from largest to smallest in distance from the mean (where "feature importance" is the reduction of triangle accuracy that results from removing the given feature on the given trial). The mean change is 0.12%, indicating a tendency for feature importance to increase modestly on average as the model gets smaller. Table 1 includes all cases where the change is further than one standard deviation (2.35%) from this mean.

We will not address every entry in this list. The sections below will instead discuss some of the more notable interactions shown.

5.1. *Removal of CT-M*

The largest two changes, and four of the largest 20, occurred between trials 12 and 13 after the removal of the CT-M feature. All of these are increases in sensitivity, which indicate redundancies. These redundancies apparently involve a large group of parameters. Most non-chord-tones will occur when the left and right notes are of the same harmonic class and the middle note has a conflicting scale degree. Therefore, the combination of HC-LR and SD-M can substitute for CT-M in many instances, making for a three-way redundancy. This is indicated in the first two entries of Table 1.

Other kinds of non-chord-tones that occur over changes of harmony, such as accented dissonances and passing tones preceding a chord change, can be predicted with the help of metrical information and interval patterns associated with these types of dissonance figures. This explains the entries 11 and 13 on Table 1, which indicate a larger five-way redundancy between CT-M, SD-M, HC-LR, Int-LMR, and Met-LMR.

These observations help us interpret the ordering of features arrived at in the experiment. The importance of chord-tone status is reflected not only in the value of CT-M in the results of Experiment 2, but also in the large increases in sensitivity of the four features in Table 1. Therefore we may conclude that chord-tone status, as might be expected, plays a large role in the analytical process for these kinds of short excerpts.

5.2. *Removal of temporal features*

Trial 14, where the Lev1-LR feature is removed, is also prominent in Table 1. All of the large changes on this trial reflect interdependencies: entries 4 (Met-LMR), 6 (HC-LR), 16 (Int-LMR), and 21 (SD-M). These are all the features remaining in the model at trial 14. This result indicates that each of these basic metrical, harmonic, and melodic features operate differently at the local, note-to-note level than at higher levels. In other words, this is strong evidence that the rules of note-to-note analysis differ in systematic ways from higher-level analytical reasoning. For the harmonic and metrical features, this is readily understandable – a change of harmony at the note-to-note level is likely to represent a direct harmonic progression, and direct metrical relations, similarly, have a more specific meaning than indirect ones. However, the inclusion of melodic

Table 1. Changes of feature importance from one trial to the next greater than one standard deviation from the mean of 0.12% (less than −2.23% or greater than 2.47%).

No.	Feature	Trial	Feature removed	Edge accuracy	Difference from full	Change from previous trial
1	SD-M	13	CT-M	52.61	13.38	12.36
2	HC-LR	13	CT-M	52.74	13.25	7.77
3	SD-M	10	SD-LR	60.00	8.15	7.39
4	Met-LMR	14	Lev1-LR	60.00	0.51	− 5.99
5	CT-M	10	SD-LR	67.52	0.64	− 5.86
6	HC-LR	14	Lev1-LR	52.99	7.52	− 5.73
7	Int-LMR	10	SD-LR	59.36	8.79	5.73
8	Int-LMR	15	Met-LMR	45.73	14.27	5.22
9	MetN-LR	10	SD-LR	68.28	− 0.13	− 4.71
10	SD-M	11	MetN-LR	64.84	3.44	− 4.71
11	Int-LMR	13	CT-M	53.50	12.48	4.59
12	Int-LMR	12	Lev2-LR	58.85	7.90	− 4.08
13	Met-LMR	13	CT-M	59.49	6.50	4.20
14	Int-LMR	8	RN-LR	66.62	3.95	− 3.69
15	Met-LMR	11	MetN-LR	64.20	4.08	3.82
16	Int-LMR	14	Lev1-LR	51.46	9.04	− 3.44
17	Int-LMR	5	IntD-LR	65.35	7.64	3.44
18	SD-M	16	HC-LR	46.62	5.35	− 3.06
19	Met-LMR	10	SD-LR	67.90	0.25	− 3.06
20	Int-LMR	11	MetN-LR	56.31	11.97	3.18
21	SD-M	14	Lev1-LR	50.06	10.45	− 2.93
22	RN-LR	4	HC-M	72.36	0.76	− 2.68
23	SD-LR	9	RN-M	68.15	0.51	− 2.68
24	RN-LR	3	CT-LR	69.43	3.44	2.68
25	HC-LR	10	SD-LR	63.31	4.84	2.68
26	RN-M	8	RN-LR	68.66	1.91	− 2.42
27	SD-M	12	Lev2-LR	65.73	1.02	− 2.42

features also shows that we should expect different melodic shapes at slightly higher levels than at the note-to-note level.

The latter point may also be made about Lev2-LR, the distinction between within-measure and between-measure levels. The melodic features are also dependent on this feature as indicated in entries 12 (Int-LMR) and 27 (SD-M).

5.3. *Importance of Int-LMR*

The Int-LMR feature occurs frequently in Table 1. This data is perhaps most readily assessed by considering the overall trajectory of Int-LMR's feature importance, as shown in Figure 14. There is a general upward trajectory indicating that, as features are removed, Int-LMR compensates for many of them, especially melodic and metrical features. This trend is interrupted by three prominent dips. The first corresponds to the removal of two harmonic features, suggesting that the likelihood of certain interval patterns depends upon harmonic context (e.g. a change of harmony versus a stable harmony). The other two dips are for the temporal features just discussed.

5.4. *Interactions involving SD-LR*

As previously noted, the SD-LR feature is redundant with the combination of SD-M and Int-LMR, in the sense that the two middle features may be used to predict SD-LR, with the only exception being distinctions between chromatic and diatonic values. It is therefore unsurprising that two of the strongest interactions involve redundancies between these three features (entries

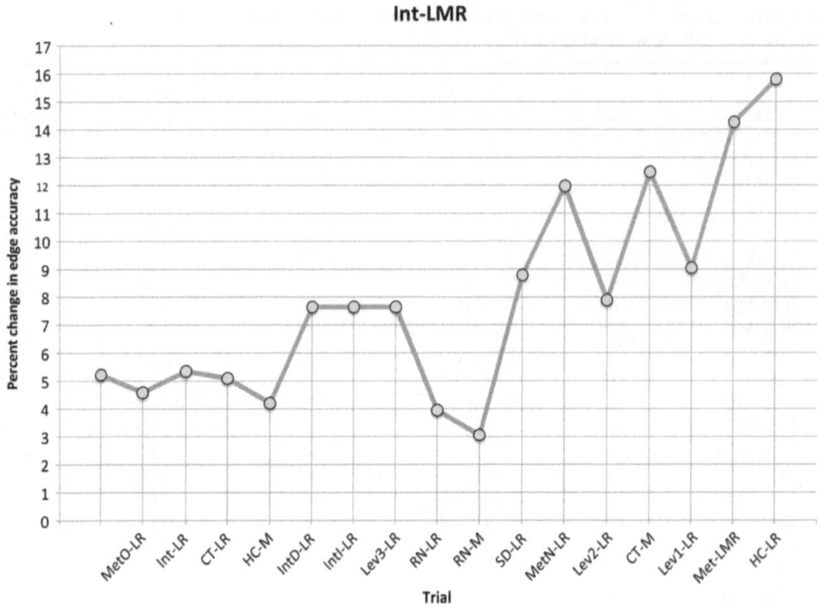

Figure 14. Changes in Int-LMR's feature importance through the trials. The labels on the horizontal axis show the feature removed prior to the given trial.

3 and 7 in Table 1). However, SD-LR also shows some prominent interdependencies indicated by negative shifts in entries 5 (CT-M), 9 (MetN-LR), and 19 (Met-LMR). These reflect the role of SD-LR as the only LR melodic feature remaining in the model beyond trial 6, which allows melodic considerations to appear on the left side of the conditional probability, as predictors. The results show that the use of melodic information to reliably predict features of the middle note is largely dependent upon complete harmonic (CT-M) and metrical (MetN-LR and Met-LMR) information.

6. Conclusions

- Melodic, harmonic, and metrical features are all significant considerations in Schenkerian analysis.
- As a whole, harmonic information is the most essential for making an accurate analysis.
- However, harmonic information breaks down into two separately important components. First, harmonic context is the most useful conditional feature for predicting characteristic patterns of all features. Second, identification of non-chord-tones is a significant aspect of the analytical process.
- Melodic information – scale-degree progression and intervallic pattern – is also crucial and actually becomes the most important analytical factor when harmonic features are divided up between harmonic context and chord tone status.
- Metrical information is the least important overall, but nonetheless does play an essential role.
- The rules of Schenkerian analysis vary from level to level. In particular, there is strong evidence that the note-to-note level follows substantially different rules than larger-scale levels. This is especially true of how melodic patterns work in Schenkerian analysis, which varies not only at the note-to-note level, but also between within-measure and across-measure levels.

Acknowledgments

The authors would like to thank the anonymous reviewers, José M. Iñesta (a Guest Editor of this special issue), and Thomas Fiore (the journal's co-Editor-in-Chief) for their helpful and insightful comments on earlier versions of this paper.

Disclosure statement

The authors have no conflict of interest.

References

Agawu, Kofi. 2009. *Music as Discourse: Semiotic Adventures in Romantic Music*. New York: Oxford University Press.

Breiman, Leo. 2001. "Random Forests." *Machine Learning* 45 (1): 5–32.

Burstein, L. Poundie. 2011. "Schenkerian Analysis and Occam's Razor." *Res Musica* 3 (3): 112–121.

Cadwallader, Allen, and David Gagné. 1998. *Analysis of Tonal Music: A Schenkerian Approach*. Oxford: Oxford University Press.

Forte, Allen, and Steven E. Gilbert. 1982a. *Instructor's Manual for "Introduction to Schenkerian Analysis"*. New York: W. W. Norton.

Forte, Allen, and Steven E. Gilbert. 1982b. *Introduction to Schenkerian Analysis*. New York: W. W. Norton.

Frankel, R. E., S. J. Rosenschein, and S. W. Smoliar. 1978. "Schenker's Theory of Tonal Music – Its Explication Through Computational Processes." *International Journal of Man–Machine Studies* 10 (2): 121–138.

Gilbert, Édouard, and Darrell Conklin. 2007. "A Probabilistic Context-Free Grammar for Melodic Reduction." In *Proceedings of the International Workshop on Artificial Intelligence and Music, 20th International Joint Conference on Artificial Intelligence,* 83–94.

Hamanaka, Masatoshi, Keiji Hirata, and Satoshi Tojo. 2005. "ATTA: Automatic Time-Span Tree Analyzer Based on Extended GTTM." In *Proceedings of the Sixth International Conference on Music Information Retrieval,* 358–365.

Hamanaka, Masatoshi, Keiji Hirata, and Satoshi Tojo. 2006. "Implementing 'A Generative Theory of Tonal Music'." *Journal of New Music Research* 35 (4): 249–277.

Hamanaka, Masatoshi, Keiji Hirata, and Satoshi Tojo. 2007. "ATTA: Implementing GTTM on a Computer." In *Proceedings of the Eighth International Conference on Music Information Retrieval,* 285–286.

Hamanaka, Masatoshi, and Satoshi Tojo. 2009. "Interactive GTTM Analyzer." In *Proceedings of the 10th International Society for Music Information Retrieval Conference,* 291–296.

Hastie, Trevor, Robert Tibshirani, and Jerome Friedman. 2009. *The Elements of Statistical Learning*. 2nd ed. New York: Springer-Verlag.

Jiménez, Víctor M., and Andrés Marzal. 2000. "Computation of the *N* Best Parse Trees for Weighted and Stochastic Context-Free Grammars." In *Advances in Pattern Recognition,* Volume 1876 of *Lecture Notes in Computer Science*, edited by Francesc J. Ferri, José M. Iñesta, Adnan Amin, and Pavel Pudil, 183–192. Berlin: Springer-Verlag.

Jurafsky, Daniel, and James H. Martin. 2009. *Speech and Language Processing: An Introduction to Natural Language Processing, Speech Recognition, and Computational Linguistics*. 2nd ed. Upper Saddle River, NJ: Prentice-Hall.

Kanamori, Kouhei, Masatoshi Hamanaka, and Junichi Hoshino. 2014. "Method to Detect GTTM Local Grouping Boundaries Based on Clustering and Statistical Learning." In *Proceedings of the 11th Sound and Music Computing/40th International Computer Music Conference,* 1193–1197.

Kassler, Michael. 1975. *Proving Musical Theorems I: The Middleground of Heinrich Schenker's Theory of Tonality*. Technical Report 103, Basser Department of Computer Science, School of Physics, University of Sydney: Australia.

Kassler, Michael. 1987. "APL Applied in Music Theory." *APL Quote Quad* 18 (2): 209–214.

Kirlin, Phillip B. 2014a. "A Data Set for Computational Studies of Schenkerian Analysis." In *Proceedings of the 15th International Society for Music Information Retrieval Conference,* 213–218.

Kirlin, Phillip B. 2014b. "A Probabilistic Model of Hierarchical Music Analysis." Ph.D. thesis, University of Massachusetts Amherst.

Kirlin, Phillip B., and David D. Jensen. 2011. "Probabilistic Modeling of Hierarchical Music Analysis." In *Proceedings of the 12th International Society for Music Information Retrieval Conference,* 393–398.

Kirlin, Phillip B., and David D. Jensen. 2015. "Using Supervised Learning to Uncover Deep Musical Structure." In *Proceedings of the 29th AAAI Conference on Artificial Intelligence,* 1770–1776.

Lerdahl, Fred, and Ray Jackendoff. 1983. *A Generative Theory of Tonal Music*. Cambridge, MA: MIT Press.

Marsden, Alan. 2005. "Towards Schenkerian Analysis by Computer: A Reductional Matrix." In *Proceedings of the International Computer Music Conference,* 247–250.

Marsden, Alan. 2010. "Schenkerian Analysis by Computer: A Proof of Concept." *Journal of New Music Research* 39 (3): 269–289.

Marsden, Alan, and Geraint A. Wiggins. 2008. "Schenkerian Reduction as Search." In *Proceedings of the Fourth Conference on Interdisciplinary Musicology*, Thessaloniki, Greece.

Mavromatis, Panayotis, and Matthew Brown. 2004. "Parsing Context-Free Grammars for Music: A Computational Model of Schenkerian Analysis." In *Proceedings of the 8th International Conference on Music Perception & Cognition*, 414–415.

Miura, Yuji, Masatoshi Hamanaka, Keiji Hirata, and Satoshi Tojo. 2009. "Use of Decision Tree to Detect GTTM Group Boundaries." In *Proceedings of the International Computer Music Conference,* 125–128.

Pankhurst, Tom. 2008. *SchenkerGUIDE: A Brief Handbook and Website for Schenkerian Analysis*. New York: Routledge.

Provost, Foster, and Pedro Domingos. 2003. "Tree Induction for Probability-Based Ranking." *Machine Learning* 52 (3): 199–215.

Rothstein, William. 1990. "The Americanization of Schenker Pedagogy?" *Journal of Music Theory Pedagogy* 4 (2): 295–300.

Schachter, Carl. 1990. "Either/Or." In *Schenker Studies*, Volume 1, edited by Hedi Siegel, 165–179. Cambridge, UK: Cambridge University Press.

Yust, Jason. 2006. "Formal Models of Prolongation." Ph.D. thesis, University of Washington, Seattle.

Yust, Jason. 2009. "The Geometry of Melodic, Harmonic, and Metrical Hierarchy." In *Proceedings of the International Conference on Mathematics and Computation in Music*, 180–192.

Yust, Jason. 2015. "Voice-Leading Transformation and Generative Theories of Tonal Structure." *Music Theory Online* 21 (4). http://www.mtosmt.org/issues/mto.15.21.4/mto.15.21.4.yust.html.

Mapping between dynamic markings and performed loudness: a machine learning approach

Katerina Kosta, Rafael Ramírez-Melendez, Oscar F. Bandtlow, and Elaine Chew

Loudness variation is one of the foremost tools for expressivity in music performance. Loudness is frequently notated as dynamic markings such as ***p*** (*piano*, meaning soft) or ***f*** (*forte*, meaning loud). While dynamic markings in music scores are important indicators of how music pieces should be interpreted, their meaning is less straightforward than it may seem, and depends highly on the context in which they appear. In this article, we investigate the relationship between dynamic markings in the score and performed loudness by applying machine learning techniques – decision trees, support vector machines, artificial neural networks, and a k-nearest neighbor method – to the prediction of loudness levels corresponding to dynamic markings, and to the classification of dynamic markings given loudness values. The methods are applied to 44 recordings of performances of Chopin's Mazurkas, each by 8 pianists. The results show that loudness values and markings can be predicted relatively well when trained across recordings of the same piece, but fail dismally when trained across the pianist's recordings of other pieces, demonstrating that score features may trump individual style when modeling loudness choices. Evidence suggests that all the features chosen for the task are relevant, and analysis of the results reveals the forms (such as the return of the theme) and structures (such as dynamic-marking repetitions) that influence the predictability of loudness and markings. Modeling of loudness trends in expressive performance appears to be a delicate matter, and sometimes loudness expression can be a matter of the performer's idiosyncracy.

1. Introduction

Information based on loudness changes abstracted from music signals can serve as an important source of data for the creation of meaningful and salient high-level features of performed music, which is almost all of the music that we hear. Our focus is on concepts related to dynamic levels in musical expressivity which are represented in the score by markings such as ***p*** (*piano*, meaning soft) and ***f*** (*forte*, meaning loud). These symbols are interpreted in performance and communicated through the varying of loudness levels. Understanding the relationship between dynamic markings and loudness levels is critical to applications in music cognition, musicology (performance analysis), and music informatics (e.g. transcription).

Our approach is a computational one. Existing work on computational modeling of loudness in performed music is limited and has largely focused on expressive performance rendering. Here, loudness is considered as part of the performances characteristics that have a large impact on the quality of the rendered pieces. One example is the YQX probabilistic performance rendering system based on Bayesian network theory described by Widmer, Flossmann, and Grachten (2009). In this study, loudness is, among others, one of the expressive performance rendering targets. A similar approach for defining loudness as a parameter of expression was taken by Tae Hun et al. (2013) in their statistical modeling of polyphonic piano renditions.

Other researchers have studied dynamics as an important tool for shaping performance using computational techniques. Widmer and Goebl (2004) reviewed various uses of Langner's tempo–loudness space in the representation and shaping of expressive performance parameters. Sapp (2008) used scapeplots to represent loudness variation at different hierarchical levels and to distinguish between performance styles.

Little work exists on score-based loudness feature representation, an exception being the study of Kosta, Bandtlow, and Chew (2014) on the meanings of dynamic markings in performances of five Chopin Mazurkas. A related study is that of Grachten and Krebs (2014) where a machine learning approach is applied to score-based prediction of note intensities in performed music. From the perspective of transcribing the loudness changes to dynamic markings, the *MUDELD* algorithm proposed by Ros et al. (2016) uses linguistic description techniques to categorize the dynamic label of separate musical phrases into three levels labeled as *piano* (***p***), *mezzo* (***m***), and *forte* (***f***).

In this article, we present a number of machine learning approaches with a two-fold purpose: predicting loudness levels given dynamic score markings and classifying performed loudness into dynamic markings. This will be done by taking into account features that emphasize the fact that dynamic markings have to be understood in relative terms. The approach to the analysis of dynamic markings is inspired by the description of dynamics by Khoo (2007) as operating on "primary dynamic shading" (absolute) and "inner shadings" (relative) levels. A consequence is that the absolute loudness of dynamic markings may be superseded by their local context so that a ***p*** dynamic might objectively be louder than an ***f*** at another part of the piece, as shown by Kosta, Bandtlow, and Chew (2014).

Fabian, Timmers, and Schubert (2014) posit that while a significant part of the performer's aim is to communicate the composer's intentions, nevertheless, performers bring their personal, cultural, and historical viewpoints to the fore when subjectively understanding expression. In an early study on dynamics performance styles, Repp (1999) alludes to the difficulty of grouping patterns of dynamic changes across recordings of the same musical excerpt, highlighting that only a weak relationship exists between performers' sociocultural variables with the dynamics profile in their recordings. In order to systematically investigate variability in the performer's understanding of the composer's intentions, we analyse eight different pianists' recordings of 44 distinct Mazurkas by Frédéric Chopin.

The remainder of the article is organized as follows: in Section 2 we describe the dataset used for this study, the features extracted, the learning task, and the algorithms employed. Section 3 presents and discusses the results obtained when predicting loudness levels at dynamic markings; and Section 4 does the same for results obtained when classifying loudness levels into dynamic markings. Finally, Section 5 summarizes the conclusions with some general discussions.

2. Material and methods

This section describes the music material and computational methods that form the basis of the studies of this article. Section 2.1 describes the dataset, the multiple alignment strategy used to synchronize audio and score features, and the features extracted. Section 2.2 presents the machine learning algorithms employed in the experiments.

2.1. *Data preparation*

For the purpose of this study, we examine recordings of 8 pianists' performances of 44 Mazurkas by Frédéric Chopin. The Mazurkas and the number of dynamic markings are each detailed in Table 1, and the 8 pianists together with the recording's year and index are identified in Table 2. The audio data were obtained from the CHARM Project's Mazurka database.[1]

Table 1. Chopin Mazurkas used in this study and the number of dynamic markings that appear in each one. Mazurkas are indexed as "M < *opus* > - < *number* > ."

Mazurka index	M06-1	M06-2	M06-3	M07-1	M07-2	M07-3	M17-1	M17-2	M17-3	M17-4	M24-1
No. of markings	18	13	22	13	13	18	7	6	9	7	5
Mazurka index	M24-2	M24-3	M24-4	M30-1	M30-2	M30-3	M30-4	M33-1	M33-2	M33-3	M33-4
No. of markings	12	7	33	8	14	25	18	5	16	4	12
Mazurka index	M41-1	M41-2	M41-3	M41-4	M50-1	M50-2	M50-3	M56-1	M56-2	M56-3	M59-1
No. of markings	12	5	6	7	15	14	17	14	7	16	8
Mazurka index	M59-2	M59-3	M63-1	M63-3	M67-1	M67-2	M67-3	M67-4	M68-1	M68-2	M68-3
No. of markings	8	11	9	4	17	10	13	11	12	21	8

Table 2. Pianist's name, year of the recording, and pianist ID.

	Pianist							
	Chiu	Smith	Ashkenazy	Fliere	Shebanova	Kushner	Barbosa	Czerny
Year	1999	1975	1981	1977	2002	1990	1983	1989
ID	P1	P2	P3	P4	P5	P6	P7	P8

The loudness time series is extracted from each recording using the *ma_sone* function in Elias Pampalk's Music Analysis toolbox.[2] The loudness time series is expressed in sones, and smoothed by local regression using a weighted linear least squares and a 2nd degree polynomial model (the "loess" method of MATLAB®'s *smooth* function[3]). The final values are normalized by dividing each value with the largest one per recording; in this way we are able to compare different recording environments.

2.1.1. *Audio recording alignment.*

To speed the labour-intensive process of annotating beat positions for each recording, only one recording (which we refer to as the "reference recording") for a given Mazurka was annotated manually, and the beat positions transferred automatically to the remaining recordings using a multiple performance alignment heuristic to be described below. The multiple performance alignment heuristic employs the pairwise alignment algorithm by Ewert, Müller, and Grosche (2009), which is based on Dynamic Time Warping (DTW) applied to chroma features. This pairwise alignment technique extends previous synchronization methods by incorporating features that indicate onset positions for each chroma. Ewert, Müller, and Grosche report a significant increase in alignment accuracy resulting from the use of these chroma-onset features, and the average onset error for piano recordings is 44 milliseconds (ms).

[1] http://www.mazurka.org.uk, accessed 20 February 2016.
[2] http://www.pampalk.at/ma/documentation.html, accessed 20 February 2016.
[3] http://uk.mathworks.com/help/curvefit/smooth.html?refresh = true, accessed 20 February 2016.

The pairwise alignment algorithm creates a match between two audio files, say i and j, using dynamic time warping. The matched result is presented in the form of two column vectors $\mathbf{p}_i$ and $\mathbf{q}_j$, each with m entries where m depends on the two recordings chosen, i and j. Each vector presents a nonlinear warping of the chroma features for the corresponding audio file, and represents the timing difference between the two recordings. A pair of entries from the two vectors gives the indices of the matching time frames of the two audio files. We compute the Euclidean distance between each pair of the dynamic time warped audio files as follows:

$$d_{i,j} = \sqrt{\sum_{k=1}^{m}(q_{j,k} - p_{i,k})^2}, \quad \forall i \neq j, \tag{1}$$

where $m \in \mathbb{N}$ is the size of the vectors. In this way, each audio has a profile corresponding to its alignment to all other audio recordings which is $\mathbf{d}_i = [d_{i,j}]$. The average value of all the alignment accuracies for the ith recording in relation to the remaining ones is $\overline{\mathbf{d}}_i$.

The goal of the multiple performance alignment heuristic is to optimize the choice of a *reference audio* with which we can obtain better alignment accuracies than with another reference audio file. We consider the best reference file to be one that minimizes the average distance to other audio files and without extreme differences from more than two other audio recordings as measured by the norm distance. Mathematically, the problem of finding the reference audio can be expressed as one of solving the following problem:

$$\begin{aligned} &\min_i \overline{\mathbf{d}}_i \\ \text{s.t.} \quad &\#\left\{j : |d_{i,j}| > q_3(\mathbf{d}_i) + 1.5[q_3(\mathbf{d}_i) - q_1(\mathbf{d}_i)]\right\} \leq 2, \end{aligned}$$

where $q_\ell(\mathbf{d}_i)$ is the ℓth quantile of $\mathbf{d}_i$, and the left-hand side of the inequality uses an interquartile-based representation of an outlier. The "reference recording" is then given by $\arg\min_i \overline{\mathbf{d}}_i$.

We then detect manually the beat positions only for the reference audio recording and we obtain the beat positions of the remaining recordings by using the alignment method mentioned above. In order to evaluate the method, we have compared these derived beat positions with our manually annotated ones for 44 recordings of the Mazurka Op. 6 No. 2 and the average error was 37 ms.

2.1.2. *Observed dynamic features.*

The dynamic markings are taken from the Paderewski, Bronarski, and Turczynski (2011) edition of the Chopin Mazurkas. For the dataset described in Table 1, ***pp*** occurs 63 times, ***p*** 234, ***mf*** 21, ***f*** 169, and ***ff*** 43 times, giving a total of 530 dynamic markings. The loudness value corresponding to each marking is the average of that found at the beat of the marking and the two subsequent beats. More formally, if $\{y_n\} \in \mathbb{R}$ is the sequence of loudness values in sones for each score beat indexed $n \in \mathbb{N}$ in one piece, then the loudness value associated with the marking at beat b is $\ell_b = \frac{1}{3}(y_b + y_{b+1} + y_{b+2})$.

About the choice of three beats, sometimes the actual change in response to a new dynamic marking does not take place immediately, and can only be observed in the subsequent beat or two. It is clear from the data that loudness varies considerably between one dynamic marking and the next. Thus, we additionally aim to have the smallest window possible to capture dynamic changes in response to a marking. Consequently, we have chosen a three-beat window as the window for study, which for Mazurkas corresponds to a bar of music.

For each dynamic marking in the dataset, we have extracted the following associated features:

(1) label of the current dynamic marking (M);
(2) label of the previous dynamic marking (PR_M);
(3) label of the next dynamic marking (N_M);
(4) distance from the previous dynamic marking (Dist_PR);
(5) distance to the next dynamic marking (Dist_N);
(6) nearest non-dynamic marking annotation between the previous and current dynamic marking, e.g. ***crescendo*** (Annot_PR);
(7) nearest non-dynamic marking annotation between current and next dynamic marking, e.g. ***crescendo*** (Annot_N); and
(8) any qualifying annotation appearing simultaneously with the current dynamic marking, e.g. ***dolcissimo*** (Annot_M).

In addition to the feature set described above, we also have an associated loudness value, L, which is the ℓ_b value on the beat of the dynamic marking.

We consider at most one value each for the features "Annot_PR," "Annot_N," and "Annot_M." If two different annotations occur on the same beat, we choose the one related to dynamic changes, an exception being the case where the annotations ***sf*** and ***a tempo*** appear simultaneously. In this case, to be consistent with the time range of other non-dynamic marking annotations, we choose ***a tempo*** over ***sf***, as it applies to more than one score beat. In the case where there was an annotation related to change in dynamics and a qualifying term such as ***poco*** preceding it, we use the annotation without the qualifier, to limit the number of dynamic terms.

2.2. *Learning task and algorithms*

In this article we explore different machine learning techniques to induce a model for predicting the loudness level at particular points in the performance. Concretely, our objective is to induce a regression model M of the following form:

$$M\,(FeatureSet) \rightarrow Loudness,$$

where M is a function that takes as input the set of features (*FeatureSet*) described in the previous section, and *Loudness* is the predicted loudness value. In order to train M we have explored the following machine learning algorithms (as implemented in Hall et al. [2009]).

Decision Trees (DTs)

Decision trees Quinlan (1986) use a tree structure to represent possible branching on selected attributes so as to predict an outcome given some observed features. The decision tree algorithm recursively constructs a tree by selecting at each node the most relevant attribute. The selection of the most relevant attribute, at each node of the tree, is based on the *information gain* associated with each attribute and the instances at each node of the tree. For a collection of loudness values, suppose there are b instances of class B and c instances of class C. An arbitrary object will belong to class B with probability $b/(b+c)$ and to class C with probability $c/(b+c)$. The expected information needed to generate the classification for the instance is given by

$$I(b,c) = -\left(\frac{b}{b+c}\log_2\frac{b}{b+c} + \frac{c}{b+c}\log_2\frac{c}{b+c}\right). \tag{2}$$

Suppose attribute A can take on values $\{a_1, a_2, \ldots, a_v\}$ and is used for the root of the decision tree, partitioning the original dataset into v subsets, with the ith subset containing b_i objects of class B and c_i of class C. The expected information required for the ith subtree is $I(b_i, c_i)$, and the expected information required for the tree with A as root is the weighted average

$$E(A) = \sum_{i=1}^{v} \frac{b_i + c_i}{b + c} I(b_i, c_i), \tag{3}$$

where the weight for the ith subtree is the proportion of the objects in the ith subtree. The information gained by branching on A is therefore $gain(A) = I(b, c) - E(A)$, and the attribute chosen on which to branch at the next node is thus $\arg\max_A gain(A)$. Decision trees can also be used for predicting numeric quantities. In this case, the leaf nodes of the tree contain a numeric value that is the average of all the training set values. Decision trees with averaged numeric values at the leaves are called *regression trees*.

Support Vector Machines (SVMs)

The *support vector machines* of Cristianini and Shawe-Taylor (2000) aim to find the hyperplane that maximizes the distance from the nearest members of each class, called *support vectors*. Cristianini and Shawe-Taylor use a nonlinear function $\phi(\cdot)$ to map attributes to a sufficiently high dimension, so that the surface separating the data points into categories becomes a linear hyperplane. This allows the model to predict nonlinear models using linear methods. The data can be separated by a hyperplane and the support vectors are the critical boundary instances from each class.

The process of finding a maximum margin hyperplane only applies to classification. However, support vector machine algorithms have been developed also for regression problems, i.e. numeric prediction, that share many of the properties encountered in the classification case. SVM for regression problems also produce a model that can usually be expressed in terms of a few support vectors and can be applied to nonlinear problems using kernel functions.

Artificial Neural Networks (ANNs)

Artificial neural networks are generally presented as systems of interconnected "neurons" which exchange messages between each other. They are based on the idea of the *perceptron* which includes an input, a hidden, and an output layer of data points connected with nodes having numeric weights. The nodes of the input layer are passive, meaning they do not modify the data. The two aspects of the problem are to learn the structure of the network and to learn the connection weights. A *multilayer perceptron (MLP)* – or *artificial neural network (ANN)* – has a *linear activation function* in all neurons, that is, a linear function that maps the weighted inputs to the output of each neuron. More formally, we assume that in one network there are d inputs, M hidden units, and c output units. As described by Bishop (1995), the *output of the jth hidden unit* is obtained by first forming a weighted linear combination of the d input values to give

$$a_j = \sum_{i=1}^{d} w_{ji}^{(1)} x_i, \tag{4}$$

where $w_{ji}^{(1)}$ denotes a weight in the first layer, going from input i to hidden unit j. The *activation of hidden unit j* is then obtained by transforming the linear sum above using an activation function

$g(\cdot)$ to give

$$z_j = g(a_j). \tag{5}$$

The *outputs of the network* are obtained by transforming the activations of the hidden units using a second layer of processing elements. For each output unit k, we construct a linear combination of the outputs of the hidden units of the form

$$a_k = \sum_{j=1}^{M} w_{kj}^{(2)} z_j. \tag{6}$$

The *activation of the kth output unit* is then obtained by transforming this linear combination using a nonlinear activation function $\tilde{g}(\cdot)$ to give

$$y_k = \tilde{g}(a_k). \tag{7}$$

We next consider how such a network can learn a suitable mapping from a given dataset. Learning is based on the definition of a suitable *error function*, which is minimized with respect to the weights and biases in the network. If we define a network function, such as the sum-of-squares error which is a differentiable function of the network outputs, then this error is itself a differentiable function of the weights. We can therefore evaluate the derivatives of the error with respect to the weights, and these derivatives can then be used to find weight values which minimize the error function.

In this article, in order to determine the weights, that is, to tune the neural network parameters to best fit the training data, we apply the *gradient descent back propagation algorithm* of Chauvin and Rumelhart (1995). The back propagation algorithm is used for evaluating the derivatives of the error function and learns the weights for a multi-layer perceptron, given a network with a fixed set of units and interconnections. The idea behind this algorithm is that the output corresponds to a propagation of errors backwards through the network. We empirically set the *momentum* applied to the weights during updating to 0.2 and the *learning rate*, which is the number of weights that are updated, to 0.3. We use a fully-connected multi-layer neural network with one hidden layer, meaning that we have one input and one output neuron for each attribute.

k-Nearest Neighborhood (k-NN)

k-NN is a type of instance-based learning that, instead of performing explicit generalization, compares the new data instances with instances in the training set previously stored in memory. In k-NN, generalization beyond the training data is delayed until a query is made to the system. The algorithm's main parameter is k, the number of considered closest training vectors in the feature space. Given a query, the output's value is the average of the values of its k nearest neighbors. More formally, as described by Hastie, Tibshirani, and Friedman (2001), the *k-nearest neighbor fit for the prediction* $\hat{Y}$ *of the output* Y is

$$\hat{Y}(x) = \frac{1}{k} \sum_{x_i \in N_k(x)} y_i, \tag{8}$$

where $N_k(x)$ is the neighborhood of the input instance x defined by the closest points x_i in the training sample.

In this article we report the results of the k-NN algorithm with $k = 1$, which finds the training instance closest in Euclidean distance to the given test instance. If several instances are qualified as the closest, the first one found is used. We tested the values of k equal to 1, 2, and 3 and there was no significant difference between the regression results.

3. Performed loudness-level modeling

This section assesses the fit of the machine learned models as they predict loudness values in a pianist's recording of a Mazurka, first given other recordings of that Mazurka, and second given other recordings of that pianist. Two experiments have been conducted for this purpose. In the first one, each prediction model has been trained for each Mazurka separately. Then, each model has been evaluated by performing an 8-fold training-test validation in which instances of one pianist of the training set are held out in turn as test data while the instances of the remaining 7 pianists are used as training data. In the second one, each prediction model has been trained for each pianist separately. Then each model has been evaluated by performing a 44-fold training-testing validation in which instances of one Mazurka of the training set are held out in turn as test data while the instances of the remaining 43 Mazurkas are used as training data.

Section 3.1 presents the machine learning algorithms' results of the first experiment, when predicting loudness given other pianists' recordings of the target piece; Section 3.2 presents the machine learning algorithms' results of the second experiment, when predicting loudness given the target pianist's other recordings; Section 3.3 considers the extremes in prediction results for the second experiment; Section 3.4 considers the degree of similarity in approach to loudness between pianists; and, Section 3.5 considers the relevance of the features selected.

3.1. *Predicting loudness given other pianists' recordings of the target piece*

In the first experiment, we use the machine learning methods described above to predict the loudness values at the dynamic markings of one Mazurka, given the loudness levels at the markings of the same Mazurka recorded by other pianists. We test the machine learning models by assessing their predictions at points where the composer (or editor) has placed a dynamic marking because this is our closest source of ground truth. The evaluations focus on how well a pianist's loudness choices can be predicted given those of the other 7 pianists for the same Mazurka.

As a measure of accuracy, we compute the *Pearson correlation coefficient* between predicted (X) and actual (Y) loudness values using the formula

$$r = \frac{\sum_{i=1}^{n}(x_i - \overline{X})(y_i - \overline{Y})}{\sqrt{\sum_{i=1}^{n}(x_i - \overline{X})^2}\sqrt{\sum_{i=1}^{n}(y_i - \overline{Y})^2}}, \tag{9}$$

where $x_i \in X$, $y_i \in Y$, and the size of X, Y, and n, varies from one Mazurka to the next, and is given in Table 1.

For each Mazurka, we compute the mean Pearson correlation coefficients over all recordings of that Mazurka to produce the graph in Figure 1. The results of each machine learning algorithm – decision trees (DTs), support vector machines (SVMs), artificial neural networks (ANNs), and k-nearest neighbor (k-NN) – is denoted on the graph using a different symbol.

Observe in Figure 1 that, with few exceptions, the mean Pearson correlation coefficient is fairly high. When disregarding the two most obvious outliers, Mazurka Op. 17 No. 3 and Mazurka Op. 24 No. 3, the mean Pearson correlation value ranged from 0.5192 to 0.9667 over all machine learning methods. Furthermore, the mean Pearson correlation coefficient, averaged over the four machine learning techniques, and over all Mazurkas, is equal to 0.8083. This demonstrates that, for most Mazurkas, the machine learning methods, when trained on data from other pianists' recordings of the Mazurka, can reasonably predict the loudness choices of a pianist for that Mazurka.

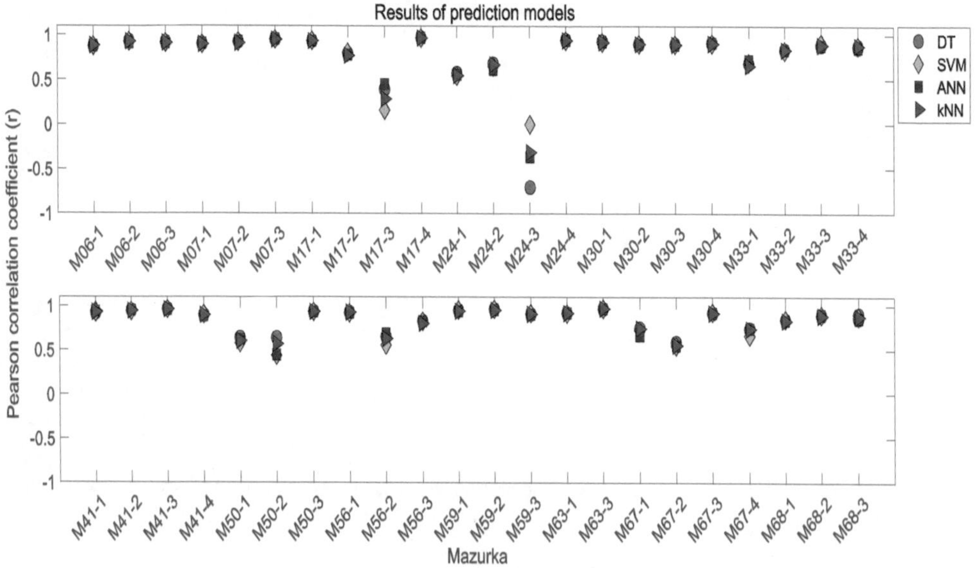

Figure 1. Pearson correlation coefficient between predicted and actual loudness values for each Mazurka, averaged over all recordings of the Mazurka, for each machine learning method – decision trees (DTs), support vector machines (SVMs), artificial neural networks (ANNs), and k-nearest neighbor (k-NN).

In the next sections, we inspect the anomalous situations of Mazurka Op. 17 No. 3 and Mazurka Op. 24 No. 3, and the special cases when one particular pianist deviates from the behavior of other pianists for specific Mazurkas.

3.1.1. *Cases of low correlation between predicted and actual loudness values*

While the overall Pearson correlation measure of success is good for the prediction of loudness values in a Mazurka recording when a machine learning model is trained on other pianists' recordings of the Mazurka, two Mazurkas were found to be outliers to this positive result: Mazurka Op. 24 No. 3 and Mazurka Op. 17 No. 3.

For Mazurka Op. 24 No. 3, the mean (over all recordings) Pearson correlation value, when averaged over the four machine learning techniques, is −0.347, meaning that the predictions are weakly negatively correlated from the actual loudness values. The mean Pearson correlation value for Mazurka Op. 17 No. 3, when averaged over the four machine learning methods, while positive, is low at 0.320, meaning that the predictions are only weakly correlated with the actual loudness values.

Looking deeper into the case of these two Mazurkas, apart from the common key of A♭ major, they also share the property of having only the dynamic markings ***p*** and ***mf*** in the score, with extended mono-symbol sequences of ***p***'s. In the case of Mazurka Op. 24 No. 3, the existing score markings are {***mf***, ***p***, ***p***, ***p***, ***p***, ***p***, ***p***}; for Mazurka Op. 17 No. 3, the score markings are {***mf***, ***p***, ***mf***, ***p***, ***p***, ***p***, ***p***, ***mf***, ***p***}. The narrow dynamic range of the notated symbols and the consecutive strings of the same symbols will almost certainly both lead to a wide range of interpretations in order to create dynamic contrast and narrative interest.

Consider the case of Mazurka Op. 24 No. 3: Figure 2 shows the actual loudness values for the eight recordings at the points of the dynamic markings. Note that, in this Mazurka, apart from the initial ***mf***, the remaining dynamic markings are uniformly ***p***. The x-axis marks the sequence of dynamic markings in the score, and the eight distinct symbols on the graphs mark the loudness values (in sones) at these points in the recording. Note the wide range of interpretation of the

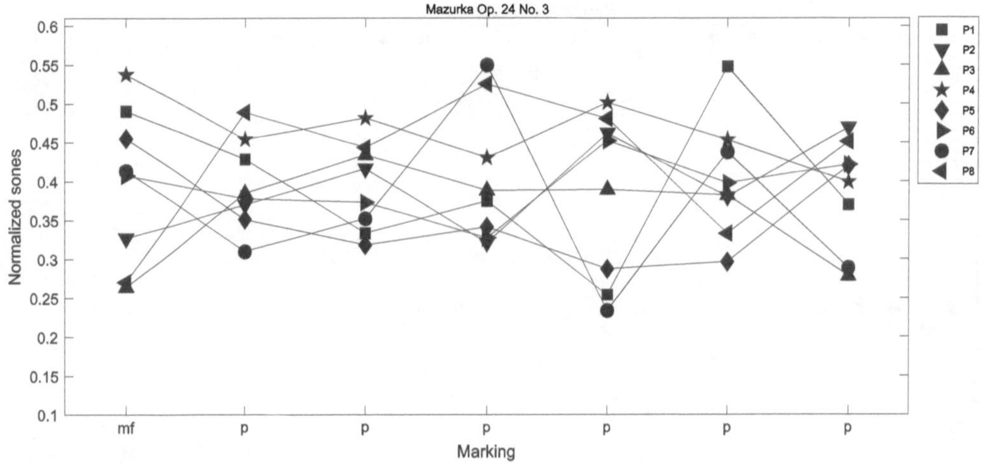

Figure 2. Representation of the loudness levels on the marking positions in score time for Mazurka Op. 24 No. 3 for the eight pianists.

loudness level for ***mf***; the loudness value ranges for the ***p***'s are often as wide as that for the ***mf***, with the recordings exhibiting many contradictory directions of change from one dynamic marking to the next. In particular, note that in three out of the eight recordings, the ***p*** immediately following the ***mf*** is actually louder than the ***mf***.

For the case of Mazurka Op. 24 No. 3, the Pearson correlation coefficient between the predicted and actual loudness values, for each of the four machine learning methods, is uniformly negative. Next, we consider the cases when the predictions are negatively correlated for only one recording of a Mazurka while the remaining seven recordings of the same Mazurka had predictions positively correlated with the actual loudness values.

3.1.2. *Cases when one recording is negatively correlated while others are not*

The tests in the previous section showed that there can be a high degree of divergence in loudness interpretation amongst recordings of a Mazurka. In this section, we examine the special cases of solitary deviant behavior, when one pianist chose starkly different loudness strategies than the others when recording a Mazurka. For this, we consider the cases when the predictions for one recording have a negative average (over all four machine learning techniques) Pearson correlation value while those for the other seven have positive average correlation coefficients between predicted and actual loudness values.

We identified four Mazurkas for which the predictions for one recording were negatively correlated on average (over all the machine learning algorithms) and the other seven were positively correlated. The average Pearson correlation values for each pianist's recordings of these four Mazurkas are plotted in Figure 3. The average correlation value of the solitary deviant recording for each Mazurka is highlighted with a circle. For each circle marking the correlation value of the worst-predicted pianist, the correlation values of the other pianists who recorded the Mazurka are shown as squares in the same vertical column.

We can see from the display that even when the machine learning algorithms did poorly in the case of a particular pianist, they often did fairly well for other pianists who recorded the same Mazurka. Figure 3 demonstrates why the loudness values of certain Mazurka recordings cannot be predicted well when machine learning algorithms are trained on other pianists' recordings of the same Mazurka.

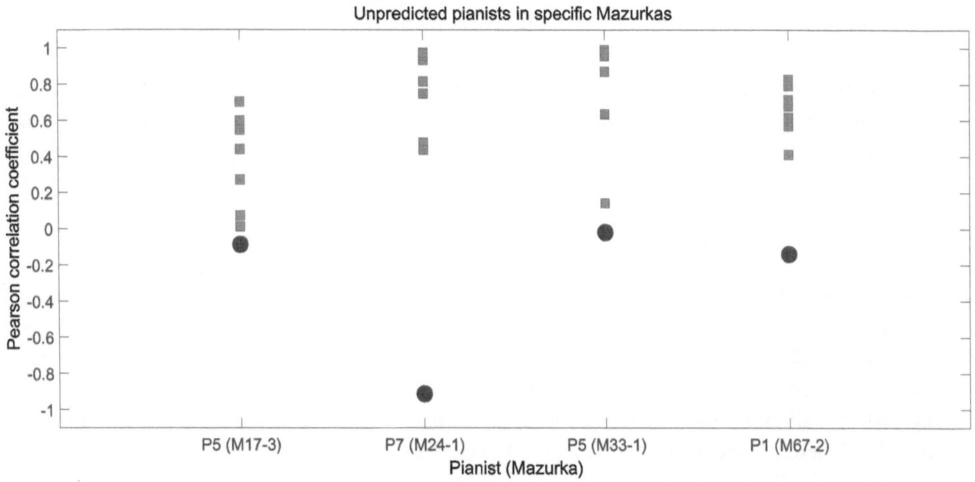

Figure 3. Pearson correlation coefficients of Mazurkas for which the worst predicted pianist's recording scored, averaging over all machine learning methods, a negative r value (circles): pianist P5 for M17-3, P7 for M24-1, P5 for M33-1, and P1 for M67-2. The average coefficients for the remaining pianists are shown as squares.

We next turn our attention to the extreme case of Mazurka Op. 24 No. 1, in which the loudness value predictions for one recording, that of Barbosa (P7), are amost perfectly negatively correlated with the actual values. Figure 4 shows the loudness time series, in score time, for all eight recordings of Mazurka Op. 24 No. 1. The loudness time series for Barbosa, the worst-predicted pianist, is highlighted in bold. The score dynamic markings for this Mazurka are labeled on the x-axis. The actual loudness values recorded for pianist P7 at these points are marked by circles. The actual loudness values of other pianists – Chiu (P1), Smith (P2), Ashkenazy (P3), Fliere (P4), Shebanova (P5), Kushner (P6), and Czerny (P8) – at this point are marked by dark symbols on the same vertical line.

As illustrated by the graph, in Barbosa's recording, he employs a performance strategy contrary to that of all or almost all of the other pianists. Furthermore, the loudness level of Barbosa's

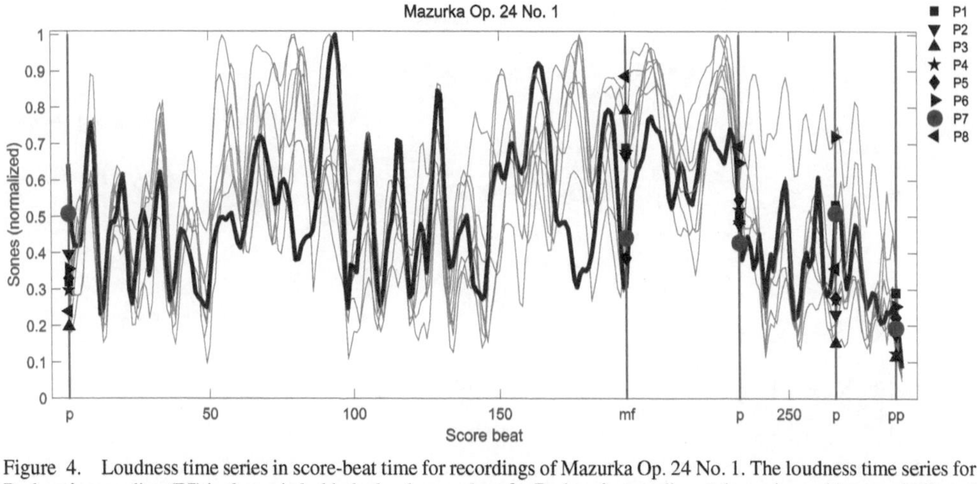

Figure 4. Loudness time series in score-beat time for recordings of Mazurka Op. 24 No. 1. The loudness time series for Barbosa's recording (P7) is shown in bold; the loudness values for Barbosa's recording at dynamic markings are indicated by circles, those of other pianists – Chiu (P1), Smith (P2), Ashkenazy (P3), Fliere (P4), Shebanova (P5), Kushner (P6), and Czerny (P8) – are shown as dark symbols.

recording sampled at the points of the first four dynamic markings are relatively level, in contrast to the strategies exhibited by the other pianists.

Having considered the prediction of loudness values in a Mazurka recording by training machine learning algorithms on recordings of the same Mazurka by other pianists, we next consider predictions of models trained on recordings of other Mazurkas by the same pianist.

3.2. *Predicting loudness given target pianist's recordings of other pieces*

In the second experiment, we use the machine learning methods to predict the loudness values at the dynamic markings of one Mazurka, given the loudness levels at the markings of other Mazurkas recorded by the same pianist. The evaluations focus on how well a pianist's loudness choices can be predicted given those made in the other 43 Mazurkas. For this purpose, a 44-fold training-testing validation has been implemented. As before, we use as the measure of prediction accuracy the Pearson correlation coefficient between actual and predicted values.

The results are displayed in Figure 5, which shows the mean, minimum, maximum, and median Pearson correlation coefficient values over all Mazurka recordings by each of the pianists listed in Table 2; the results are broken down into the four machine learning methods employed – decision trees (DTs), support vector machines (SVMs), artificial neural networks (ANNs), and k-nearest neighbor (k-NN).

Contrary to the previous experiment, where the machine learning models were trained on other pianists' recordings of the same Mazurka and did fairly well, when training the models on the same pianist's other Mazurka recordings, the average cross-validation results for each pianist are close to zero. The minimum is close to -1, implying that sometimes the loudness values of a recording can be directly contrary to the predictions, and the maximum is close to 1, implying that sometimes the loudness values behave as predicted. The results thus demonstrate that it can be extremely difficult to predict loudness values in a recording given the pianist's past behavior in recordings of other pieces. The results fare far better when training on other pianists' recordings of the same piece.

In Section 3.3, we seek to gain some insights into why the Pearson correlation values were so highly variable between the predicted and actual values for this experiment. In particular, we examine in detail the Mazurkas for which predicted and actual loudness values were most

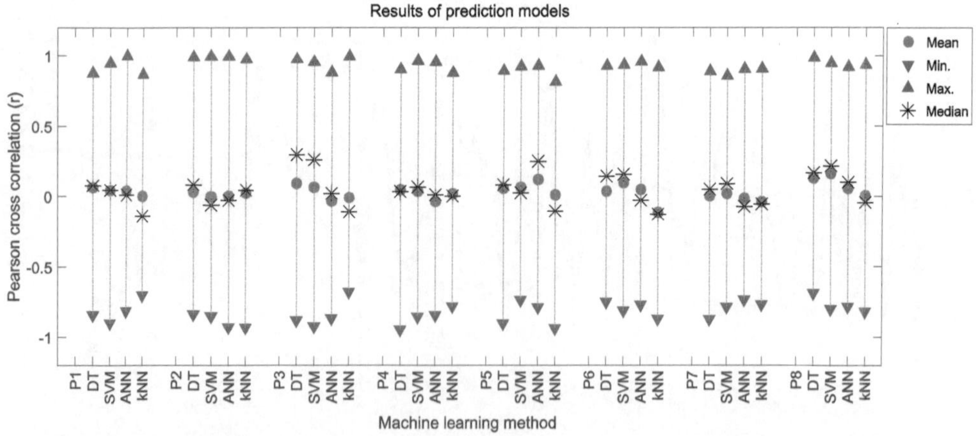

Figure 5. Pearson correlation coefficient mean, minimum, maximum, and median values for each method – decision trees (DTs), support vector machines (SVMs), artificial neural networks (ANNs), and k-nearest neighbor (k-NN) – for each pianist – Chiu (P1), Smith (P2), Ashkenazy (P3), Fliere (P4), Shebanova (P5), Kushner (P6), Barbosa (P7), and Czerny (P8).

strongly and positively correlated and most strongly and negatively correlated across all machine learning methods.

The previous results have shown that while there may be some small variability in the prediction quality of the four machine learning methods, they agree on the prediction difficulty amongst the recordings. In the next section, we perform a second check on the prediction quality using a Euclidean measure.

3.2.1. *Comparing machine learning methods*

To check for variability in the prediction quality of the four machine learning algorithms, we compute the accuracy for each algorithm using the Euclidean distance. The Euclidean distance between the predicted (X) and actual (Y) loudness values (in sones) is given by the formula

$$d(X, Y) = \sqrt{\sum_{i=1}^{n} (x_i - y_i)^2}, \tag{10}$$

where $x_i \in X$, $y_i \in Y$, and n is the size of X and Y, which varies from one Mazurka to the next, as shown in Table 1.

We average these Euclidean distances over all Mazurkas for a given pianist to produce the results shown in Figure 6. The results are grouped by pianist: Chiu (P1), Smith (P2), Ashkenazy (P3), Fliere (P4), Shebanova (P5), Kushner (P6), Barbosa (P7), and Czerny (P8). For each pianist, the graph shows accuracy results for each of the four machine learning methods – decision trees (DTs), support vector machines (SVMs), artificial neural networks (ANNs), and k-nearest neighbor (k-NN) – averaged over all Mazurka recordings by that pianist.

The results show that the algorithms perform consistently one relative to another. The span of average Euclidean distance between predicted and actual values is relatively small. In comparison, the DT algorithm produced the best results, followed closely by the SVM algorithm then the k-NN algorithm; the ANN algorithm is a more distant fourth. While the ANN algorithm fared worse in this exercise, we shall see in Section 4.2 that the ANN gives better results when the machine learning algorithms are applied to the problem of predicting dynamic-marking labels.

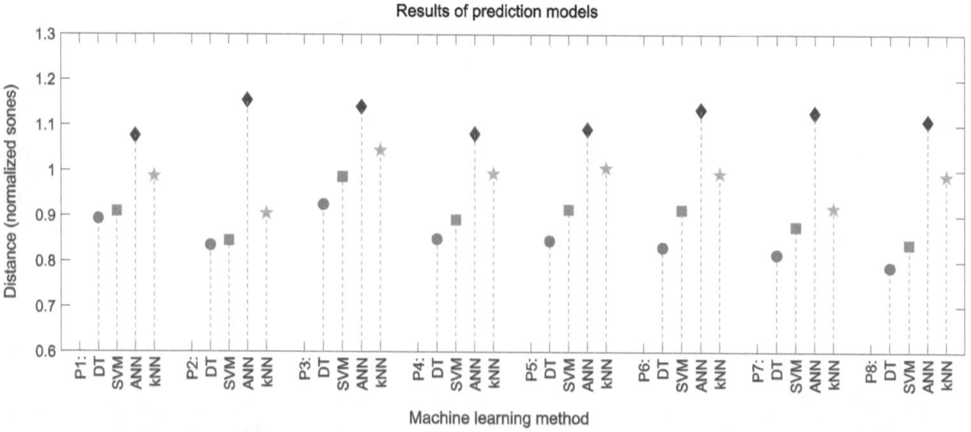

Figure 6. Euclidean distance between predicted and actual loudness values for each pianist – Chiu (P1), Smith (P2), Ashkenazy (P3), Fliere (P4), Shebanova (P5), Kushner (P6), Barbosa (P7), and Czerny (P8) – for each method – decision trees (DTs), support vector machines (SVMs), artificial neural networks (ANNs), and k-nearest neighbor (k-NN) – averaged over all Mazurka recordings by the pianist.

3.3. *A pianist's interpretation may not be predictable based on their approach to other pieces*

In this section, we dig deeper into the Pearson correlation results of the previous section to consider the extreme cases of when the predicted and actual loudness values are most consistently positively and consistently negatively correlated. We consider the Mazurkas for which all methods produced loudness predictions that were negatively correlated with the actual loudness values for all pianists; we also single out the Mazurkas for which all methods produced predictions that were positively correlated with the actual values for all pianists.

3.3.1. *Most negatively and most positively correlated results*

In the cross-validation, the highest Pearson correlation coefficient between predicted and actual loudness, 0.9981, is encountered in Chiu (P1)'s recording of Mazurka Op. 63 No. 3 (M63-3), and the lowest correlation value, −0.9444, in Fliere (P4)'s recording of Mazurka Op. 17 No. 2 (M17-2).

The four Mazurkas for which the Pearson correlation is negative for all the machine learning methods and for all eight pianists are Mazurkas Op. 7 No. 1 (M07-1), Op. 24 No. 2 (M24-2), Op. 24 No. 4 (M24-4), and Op. 50 No. 3 (M50-3). This means that, for these Mazurkas, the pianists' recorded loudness strategies are contrary to those gleaned from their recordings of the other Mazurkas. The three Mazurkas for which the Pearson correlation coefficient over all pianists and for all machine learning methods was positive are Mazurkas Op. 30 No. 4 (M30-4), Op. 41 No. 2 (M41-2), and Op. 68 No. 2 (M68-2). For these Mazurkas, the pianists' recorded loudness strategies are in accordance with those gleaned from their recordings of the other Mazurkas.

The results are summarized in Figure 7, which presents the Pearson correlation result for each of the eight pianists for the Mazurkas mentioned; each pianist's data point for a Mazurka shows the average over the four machine learning methods. Note that, for each Mazurka, the correlation coefficients are relatively closely grouped for all pianists. In the next section, we shall examine more closely the four Mazurkas having all-negative correlation values, i.e. the ones to the left of the dividing line.

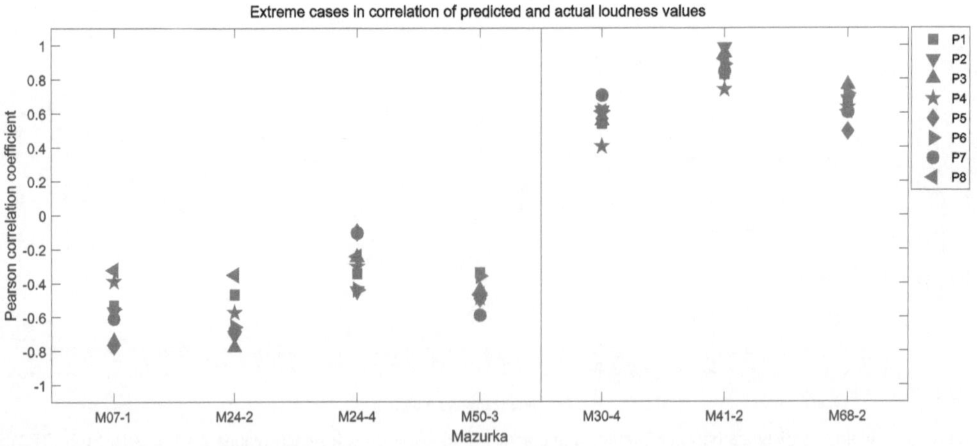

Figure 7. Pearson correlation coefficient values for each pianist, averaged over all machine learning methods, for the Mazurkas having all negative (left: M07-1, M24-2, M24-4, M50-3) and all positive (right: M30-4, M41-2, M68-2) correlation values.

3.3.2. *Cases of negative correlation*

Here, we focus on the four Mazurkas with negative Pearson correlation coefficients for all pianists and all machine learning methods. Figure 8 shows the predicted and actual loudness values at each dynamic marking in the four Mazurkas, concatenated in sequence. The values shown are the average over all pianists and all machine learning methods for each marking and Mazurka.

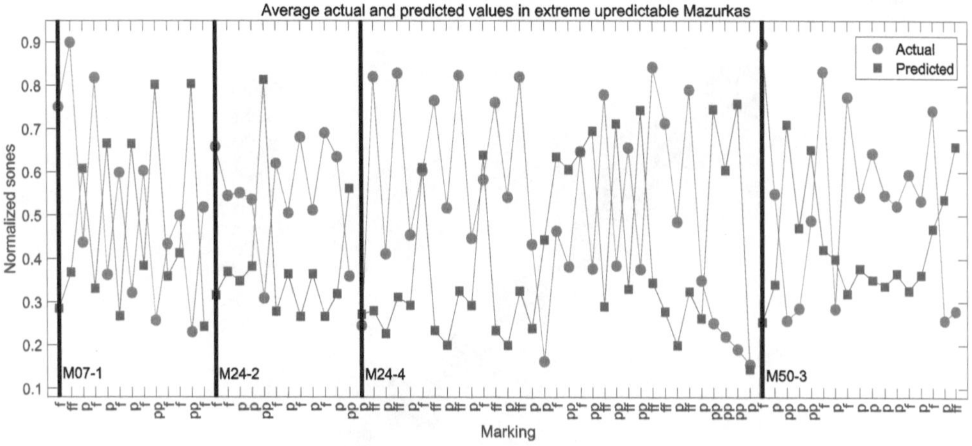

Figure 8. Average actual dynamic values (circles) and predicted dynamic values (squares) at dynamic-marking positions for Mazurkas M07-1, M24-2, M24-4, and M50-3, where the average of the Pearson correlation coefficient over pianists and methods was negative.

As can be seen, for most dynamic markings in these four Mazurkas, the actual loudness levels are the opposite of the predicted loudness values. The difference in loudness strategy may be due to deliberate contrarian behavior, or to aspects of the piece relevant to dynamic choices not being captured in the selected features. A discussion on future directions for feature analysis will be presented in Section 3.5.

There may also be score-inherent factors in these four Mazurkas that contributed to the negative correlation results. These include the oscillatory nature – and hence longer range dependencies – of some of the dynamic-marking sequences, and the presence of sequences of identical markings – which lead to greater variability in interpretation. Instances of oscillatory sequences include the sequence (***p***, ***f***, ***p***, ***f***, ***p***, ***f***) in Mazurka Op. 7 No. 1, the sequence (***f***, ***p***, ***f***, ***p***, ***f***, ***p***) in Mazurka Op. 24 No. 2, and the sequence (***pp***, ***ff***, ***pp***, ***ff***, ***pp***, ***ff***) in Mazurka Op. 24 No. 4. Examples of monosymbol sequences include the sequence of three ***pp***'s in Mazurka Op. 24 No. 4, and the sequence of four ***p***'s in Mazurka Op. 50 No. 3.

The analysis of the results of the loudness value prediction experiments leads us to suspect that the poor predictability of some of the Mazurka recordings may be due to high variability in performed loudness strategies among the pianists for the range of Mazurkas represented. Hence, in the next section, we describe and report on an experiment to test the degree of similarity in the loudness strategies employed by one pianist versus that used by another.

3.4. ***Inter pianist similarity***

To determine the degree of similarity in the loudness strategies employed by different pianists, machine learning models were trained on all Mazurka recordings by one pianist and used to predict the loudness values in all Mazurka recordings by another pianist. The goodness of fit is measured using the mean (over all Mazurka recordings) of the average (over all machine learning methods) Pearson correlation coefficient.

The results are shown in Figure 9 in the form of a matrix, where the (*i*, *j*)th element in the matrix represents the percentage of the mean averaged Pearson correlation coefficient between the loudness value predictions of the model trained on data of pianist *i* and the corresponding actual loudness data of pianist *j*. The correlation values range from 68.36 to 78.43%. This shows the high level of similarity between pianists for interpreting dynamic markings in the pieces investigated.

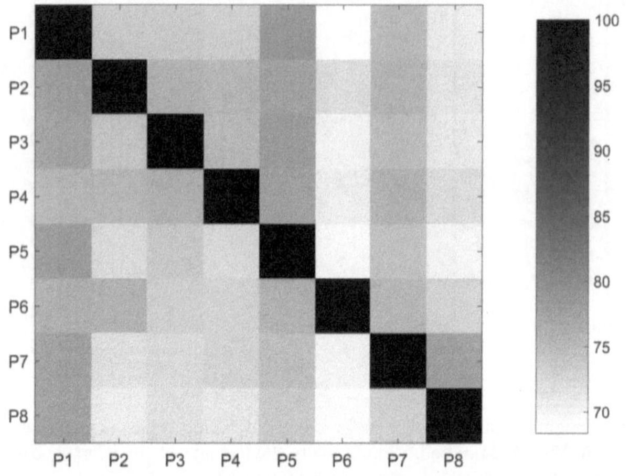

Figure 9. Matrix displaying Pearson correlation coefficient values (in %), averaged over all machine learning methods, when applying a model trained on recordings by one pianist to predict the loudness values of another pianist.

Higher correlation values are observed in columns P1 and P5. We can deduce that Chiu (P1)'s interpretation model, followed closely by Shebanova (P5) 's model, best predicts the loudness values of recordings by other pianists. Note that Chiu's recordings also achieved some of the highest correlation values, meaning that they were most predictable. Furthermore, the loudness strategies of pianists P1 and P5 best fit each other, i.e. the highest non-diagonal Pearson correlation value is found when P1's model is used to predict the loudness levels in P5's recording, and vice versa. On the other hand, the loudness strategies of pianist P6 is the one most dissimilar to that in other recordings, as shown by the almost white non-diagonal squares in column P6.

3.5. *Discussion on feature analysis*

The k-NN method is known to be highly sensitive to irrelevant features, i.e. it performs considerably less well than other algorithms in the presence of irrelevant features. As the results show no demonstrable trend in this respect, this leads us to think that all the extracted features in our feature set are indeed relevant.

A natural question that follows is: which of the extracted features are more salient for predicting performed loudness? To investigate this question, we apply the RELIEF algorithm from Robnik-Sikonja and Kononenko (1997) to rank the features according to relevance for predicting loudness. The RELIEF algorithm evaluates the worth of an attribute by repeatedly sampling an instance and considering the value of the given attribute for the nearest instance of the same and different class.

The results are shown in Table 3, with features being ranked from most important (1) to least important (8). The feature-relevance analysis indicates that the current dynamic marking proved to be the most informative feature for the task, followed by the text qualifiers, and the preceding dynamic marking. Interestingly, the ranking of features according to relevance for predicting loudness is the same for all pianists, which may indicate that all the considered pianists give the same relative importance to the investigated features when deciding loudness levels in their performances.

Table 3. Ranking of importance of the features for the loudness prediction task.

Ranking	1	2	3	4	5	6	7	8
Feature	M	Annot_M	Annot_PR	N_M	Dist_PR	PR_M	Dist_N	Annot_N

In order to evaluate the incremental contribution of each of the features studied, we have implemented the machine learning models using different subsets of features. More concretely, we have considered feature subsets by incrementally adding features to the training set one at a time, starting with the most important feature, i.e. the highest ranked, and continuing to add features according to their rank order. In Figure 10 we present the results for each pianist's recording, averaged over all machine learning methods.

As can be seen in the figure, for all pianists, the loudness prediction accuracy for their recordings, with few exceptions, increases monotonically with the number of features. This confirms our belief that all the features studied are indeed relevant and contribute to the accuracy of the loudness-level predictions for all recordings. It is worth noticing that the highest ranked feature, i.e. the dynamic marking, contains on its own substantial predictive power: the correlation coefficient of the loudness models induced with only dynamic-marking information alone ranges from 0.64 (for pianist P6's recordings) to 0.75 (for pianist P1's recordings).

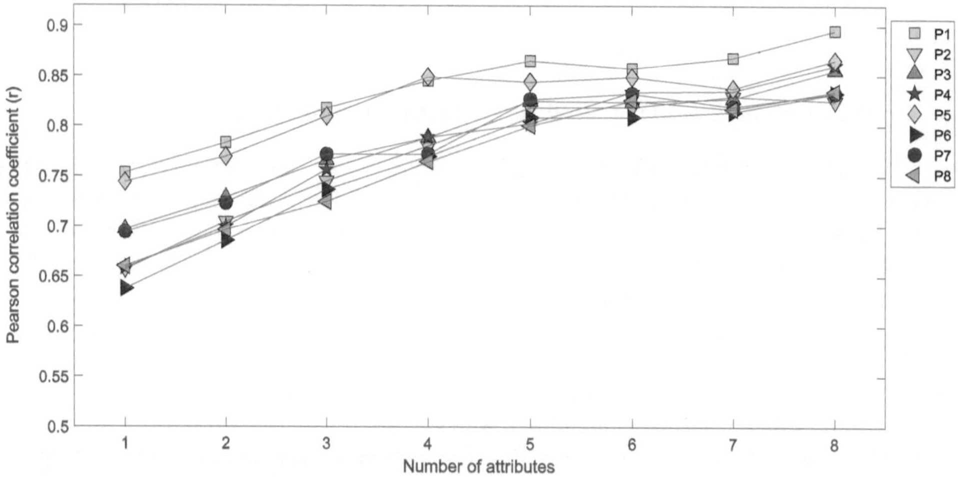

Figure 10. Upper bound for average (over all machine learning methods) Pearson correlation coefficient as number of features considered increases; features are added in the order given in Table 3. Numbers result from the optimal case where the testing data is equal to the training data.

4. Dynamic-marking prediction

Classification remains one of the staple tasks of machine learning algorithms. This section assesses the fit of the machine learned models as they are applied to the classification of a Mazurka recording's loudness values into loudness categories as indicated by dynamic markings. In particular, we examine the following problem: given a loudness level in a recorded Mazurka, what is the dynamic marking that best represents conceptually the loudness value at a particular instance in time? As in Section 3, we have conducted two experiments. In the first experiment each classification model has been trained for each Mazurka separately and an 8-fold cross-validation implemented for predicting the dynamic markings given other pianists' recordings of the target piece. In the second experiment each classification model has been trained for each pianist separately and a 44-fold cross-validation has been implemented for predicting the dynamic markings of one Mazurka given the other Mazurka recordings of the same pianist. For evaluation, we compute the percentage of correctly-classified instances.

In the remaining part of this section we present the analysis of the results of the two experiments. More specifically, Section 4.1 presents and discusses the results for predicting dynamic markings given other pianists' recordings of the target piece; Section 4.2 does the same for predicting markings given the target pianist's recordings of other pieces; Section 4.3 looks at which of the dynamic markings are more easily classified than others; and Section 4.4 evaluates the ease or difficulty of predicting the markings of a Mazurka using the ratio of correctly-classified markings.

4.1. *Predicting dynamic markings given other pianists' recordings of the target piece*

In the first experiment, we use the four machine learning methods to predict the dynamic-marking labels of loudness values given the loudness-marking data of other pianists' recordings of the target piece. The 8-fold cross-validation test results of this task give us an average of 99.22% of Correctly Classified Instances (CCIs) over all methods. This means that it is highly plausible to train a machine learning algorithm on other pianists' performances of a Mazurka to predict the dynamic-marking labels of the target recording.

We next move on to the complementary experiment in which the machine learning algorithms predict dynamic-marking labels based on loudness-marking data of other Mazurka recordings by the target pianist.

4.2. *Predicting dynamic markings given target pianist's recordings of other pieces*

In this experiment, we train classifiers – using DT, ANN, SVM, and k-NN algorithms – on the loudness-label mappings of other Mazurka recordings by the same pianist in order to identify the dynamic marking corresponding to a given loudness value in the target recording. Table 4 shows the results obtained by the classifiers in terms of the mean percentage of CCIs over all Mazurkas. The highest mean CCI values for each pianist are highlighted in bold.

As can be seen by the preponderance of numbers in bold in the ANN row, the ANN algorithm gives slightly better results in terms of mean CCI than the other methods. In particular, it yields the best results for pianists Chiu (P1), Fliere (P4), Shebanova (P5), Kushner (P6), and Czerny (P8), followed by the k-NN algorithm, which gives slightly better results for the pianists Smith (P2), Ashkenazy (P3), and Barbosa (P7). The highest average value of the percentage of correctly-classified instances over all pianists is given by the ANN algorithm. Recall that, in Section 3.2.1, ANN had the lowest prediction correlation to actual figures. That was for the case of loudness prediction; here, it performed best for dynamic-marking label prediction.

Table 4. Percentage of CCIs. Maximum values per pianist are highlighted in bold. The last column contains the average values per row and the highest average of the CCIs is highlighted in bold.

	Pianist								
Method	P1	P2	P3	P4	P5	P6	P7	P8	Average
DT	27.822	26.387	28.012	24.542	24.798	27.858	29.230	26.437	26.8858
SVM	27.735	25.283	23.962	22.830	25.283	24.717	26.227	25.670	25.2134
ANN	**30.755**	27.925	27.736	**26.981**	**26.604**	**31.132**	30.189	**30.460**	**28.9727**
k-NN	28.301	**28.491**	**28.113**	25.660	25.660	26.981	**30.377**	27.547	27.6412

The CCIs reported in Table 4 are not high, given that a classifier that always predicts ***p*** would achieve 42%. This suggests that more data are needed for this task. One avenue for further analysis is to identify the markings that are more easily classified by considering the ones that have been predicted correctly by all recordings; this study is reported in Section 4.3. Another direction is to identify the Mazurka that has the highest number of markings that are more easily predicted, by observing for every Mazurka the ratio of the markings correctly classified for all recordings to the total number of markings in the piece; this is reported in Section 4.4. We use the ANN algorithm as a basis for both studies due to its better performance found here.

4.3. *Easily predicted markings*

In this section, we seek to determine which of the dynamic-marking labels are more easily classified by the machine learning algorithms when trained on other Mazurka recordings by the target pianist. A specific marking in a Mazurka is considered to be easily predicted if it has been classified correctly for all recordings. In Table 5 we present the markings across all Mazurkas that have been identified as being easily predicted according to this criterion.

Two patterns emerge from Table 5: the first has to do with the range of the dynamic markings in the Mazurka; the second has to do with the existence of important structural boundaries such as key modulations or cadences near the marking that impact the dynamics. We shall describe each case in greater detail providing specific examples of each case.

In the first case, when considering the range of markings present in a score, the marking at the edges of this range tend to correspond to extreme loudness values in the recording. Thus, these dynamics at the extremes would be the ones most easily categorized. For example, the ***ff*** marking in Mazurka Op. 17 No. 4 (M17-4) is a case in point. The markings that appear in this Mazurka are: {***pp***,***p***,***ff***}. Clearly, ***ff*** is at the extreme of this marking set, and in the loudness spectrum. Even in its position in the score, it is placed uniquely in such a way as to highlight the extreme nature of its dynamic level. Figure 11 shows the score position of the marking: it is preceded by a ***crescendo*** and followed by a ***p***. It is no wonder that this marking is correctly classified in all recordings of this Mazurka.

Table 5. Markings that have been predicted correctly for all recordings of the Mazurka containing that marking; the numbers in parentheses indicate the position of that marking in the sequence of dynamic markings in that Mazurka.

Mazurka	Marking (position)	Mazurka	Marking (position)	Mazurka	Marking (position)
M06-3	***p*** (1), ***p*** (16)	M33-2	***ff*** (5)	M63-1	***p*** (3), ***p*** (5)
M07-3	***f*** (5)	M50-2	***p*** (2), ***p*** (4), ***p*** (6), ***p*** (7)	M67-1	***p*** (4), ***p*** (15)
M17-3	***p*** (5), ***p*** (9)	M50-3	***p*** (12)	M67-3	***p*** (11)
M17-4	***ff*** (5)	M56-1	***mf*** (12)	M68-2	***pp*** (2)
M24-4	***f*** (6), ***f*** (11)	M56-3	***p*** (12)	M68-3	***p*** (2)
M30-3	***pp*** (14)	M59-3	***p*** (11)		

Figure 11. Case of the *ff* marking in Mazurka Op. 17 No. 4 (M17-4), correctly classified in all recordings of the Mazurka when the machine learning algorithms are trained on the markings in the remaining Mazurkas by the target pianist.

Figure 12. Case of the *f* marking in Mazurka Op. 24 No. 4 (M24-4), correctly classified in all recordings of the Mazurka when the machine learning algorithms are trained on the markings in the remaining Mazurkas by the target pianist.

Figure 13. Case of the *pp* marking in Mazurka Op. 30 No. 3 (M30-3), correctly classified in all recordings of the Mazurka when the machine learning algorithms are trained on the markings in the remaining Mazurkas by the target pianist.

In the second case, structural boundaries are often inflected in performance in such a way as to reinforce their existence: the dynamic may drop immediately preceding or immediately after the boundary. As an example of a dynamic marking following a key change, consider the second *f* marking of Mazurka Op. 24 No. 4 (M24-4), which is in the key of B♭ minor. The score segment for this example is shown in Figure 12. Just before this easily predicted marking a phrase is repeated ending in the key of F major, and the repeat is marked ***sotto voce*** (in an undertone). The *f* appears at the return of B♭ minor; the performer is thus encouraged to make a sharp distinction between the *f* and the preceding ***sotto voce***. Four bars after the *f* is a ***pp*** marking, which would bring into sharper relief the *f*. These factors all conspire to make this *f* more easily detectable.

An example of an extreme dynamic marking near a structural boundary is the case of the marking ***pp*** in Mazurka Op. 30 No. 3 (M30-3), which is located at a cadence prior to a return to the main theme of the piece. The score segment for this example is shown in Figure 13. As can be seen in the score, the ***pp*** is proceeded by the text indicator ***dim.***, for diminuendo; furthermore, the text indicator ***slentando***, meaning to become slower, is paired with the marking; and, the marking is followed by an *f* marking paired with the text indicator ***risoluto***, meaning bold, at the return of the main theme. The extra meaning imputed to this ***pp*** as a signifier of the impending return of the ***f*** main theme makes it again a more extreme kind of ***pp***, and thus easier to classify.

In the next section, we consider the ease or difficulty of classifying dynamic markings through a different study, this time on the ratio of correctly-classified markings.

4.4. *Easy/hard to predict Mazurkas: ratio of correctly-classified markings*

We have observed over the course of the experiments that there were Mazurkas for which the ratio of the markings that have been correctly classified is high for each recording, while for others that same ratio is low. To study this ratio across all Mazurkas, we have run a set of cross-validation experiments for which the markings of a specific Mazurka constituted the testing set and the markings of the remaining ones constituted the training set. The resulting ratio, averaged over all recordings of the Mazurka, of correctly-classified markings to total markings, is laid out in Figure 14 in a monotonically increasing order.

Note that Mazurka Op. 24 No. 1 has the lowest average ratio. Recall that Mazurka Op. 24 No. 1 was also the one in which the loudness value predictions for one recording were almost perfectly and negatively correlated with the actual values; this was described in Section 3.1.2.

In contrast, Mazurka Op. 50 No. 2 has the highest ratio, 0.5893, meaning that this Mazurka has the highest number of correctly-classified markings for every recording. We then consider

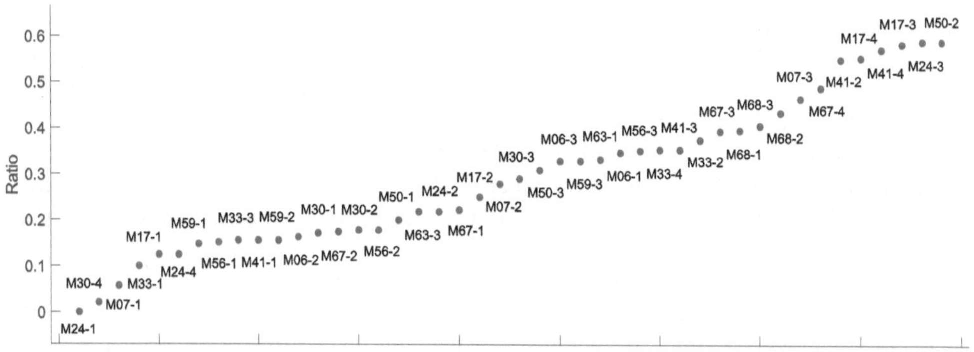

Figure 14. Average ratio, over all machine learning methods, of correctly-classified markings and number of markings per Mazurka over all pianists.

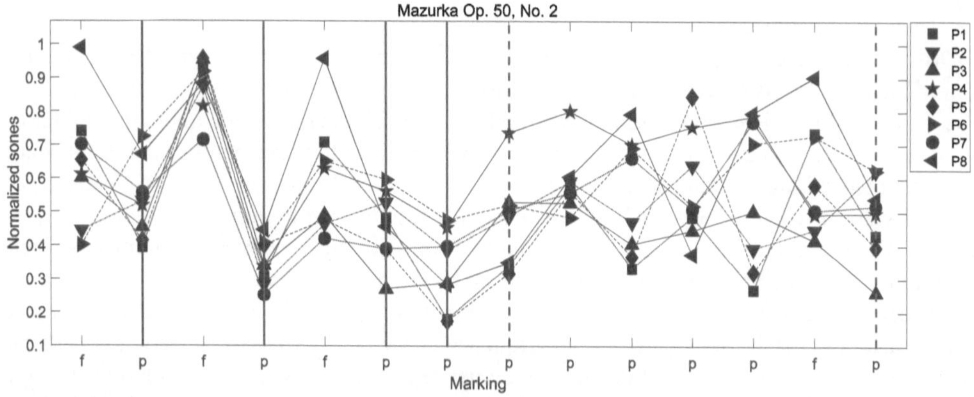

Figure 15. Loudness values at dynamic-marking positions for Mazurka Op. 50 No. 2. Solid vertical lines indicate markings that are correctly classified for all eight recordings; dashed vertical lines indicate markings that are correctly classified for seven out of eight recordings.

in detail the loudness values at dynamic markings in Mazurka Op. 50 No. 2 for all 8 recordings of that Mazurka. In Figure 15 the markings that are correctly classified for all recordings are highlighted with a solid vertical line, while the ones that are correctly classified for 7 out of 8 recordings are highlighted with a dashed vertical line. The loudness levels at these easily-classified markings follow the patterns established by other recordings.

Recall that consecutive strings of the same symbol posed significant challenges to the prediction of loudness values described in earlier sections. Note that Mazurka Op. 50 No. 2 again features a series of 7 ***p***'s, with high degrees of variation in the loudness levels at these markings and in the ***f*** immediately following the sequence.

5. Conclusions

In this article we have investigated the relationship between loudness levels and dynamic markings in the score. To this end, we have implemented machine learning approaches for the prediction of loudness values corresponding to different dynamic markings and musical contexts, as well as for the prediction of dynamic markings corresponding to different loudness levels and musical contexts. The methods – decision trees, support vector machines, artificial neural networks, and a k-nearest neighbor algorithms – are applied to 44 recordings of performances of Chopin's Mazurkas, each by 8 pianists.

The results in Section 3 highlight that, using any of the methods, loudness values and markings can be predicted fairly well when training across recordings of the same piece, but fail dismally when training across recordings of other pieces by the same pianist. This happens possibly because the score is a greater influence on the performance choices than the performer's individual style for this test set. Moreover, certain Mazurkas' loudness values can be predicted more easily than others; this is due to the range of the different markings that appear in these pieces, as well as their position and function in the score in relation to structurally important elements.

The results in Section 4 highlight the good performance of the methods in predicting the dynamic markings given the loudness-marking data of the pianists' recordings of a target piece. When training across recordings of other pieces by the same pianist, the results, while not exceptional with respect to the prediction of dynamic markings, show notable trends in markings that are classified correctly, and in pieces that have higher ratios of classified markings over all markings. These trends are based mostly on the relationship between the position of the markings and the structure of the piece.

Different tuning of the parameters in the existing machine learning techniques for the prediction of loudness levels as well as for the prediction of dynamic markings may give better results. The scope of this article, however, is not to find the optimal solution to the tuning issue, but to point out universal characteristics of this kind of complex data. Improvements of the results may occur by considering possible alterations in the feature set or the dataset. From the features set point of view, it would be especially interesting if the results were to improve with more score-informed features, and features that correspond to changes of other expressive parameters, such as tempo variations. From the dataset point of view, the poor results of the 44-fold-validation experiments suggest that there may be insufficient data or features to train a model for such a complicated task. The results may be a consequence of the choice of training data, and perhaps training on a selection of pieces that include the desired or similar sequence of markings could enhance the results. Alternatively, it may be that musicians approach each new piece differently, and that there is little benefit from training a model on recordings of other pieces to predict performance features in a target piece.

Modeling expressive performance trends using artificial intelligence appears to be a rather delicate matter. The models are limited by the particular representations of changes that happen throughout a music piece with respect to the expression rendering. The results that are drawn from the current study are likely to appear in similar studies for other instruments and datasets. Future work in this direction should include the comparison of performances from different eras, or performances of musicians from particular schools. At the same time it should be kept in mind that the freedom of employing a particular dynamic range in one interpretation is a matter of the performer's idiosyncrasy.

Acknowledgments

The authors would like to thank the anonymous referees, and also Professor Thomas Fiore, Editor-in-Chief, for their immensely helpful suggestions and comments that substantially strengthened the original submission.

Disclosure statement

No potential conflict of interest was reported by the authors.

Funding

This work was supported in part by a UK EPSRC Platform Grant for Digital Music [EP/K009559/1]; the Spanish TIN project TIMUL [TIN2013-48152-C2-2-R]; and the European Union's Horizon 2020 research and innovation programme [grant agreement No. 688269].

References

Bishop, Christopher M. 1995. *Neural Networks for Pattern Recognition*. Oxford, UK: Oxford University Press.

Chauvin, Yves, and David E. Rumelhart. 1995. *Back Propagation: Theory, Architectures, and Applications*. Hove, UK: Psychology Press.

Cristianini, Nello, and John Shawe-Taylor. 2000. *An Introduction to Support Vector Machines and Other Kernel-Based Learning Methods*. New York: Cambridge University Press.

Ewert, Sebastian, Meinard Müller, and Peter Grosche. 2009. "High Resolution Audio Synchronization Using Chroma Onset Features." In *Thirty-Fourth IEEE International Conference on Acoustics, Speech, and Signal Processing (ICASSP)*, 19–24 April 2009, Taipei, Taiwan, 1869–1872. http://dx.doi.org/10.1109/ICASSP.2009.4959972.

Fabian, Dorottya, Renee Timmers, and Emery Schubert. 2014. *Expressiveness in Music Performance: Empirical Approaches Across Styles and Cultures*. Oxford, UK: Oxford University Press.

Grachten, Maarten, and Florian Krebs. 2014. "An Assessment of Learned Score Features for Modeling Expressive Dynamics in Music." *IEEE Transactions on Multimedia* 16 (5): 1211–1218.

Hall, Mark, Eibe Frank, Geoffrey Holmes, Bernhard Pfahringer, Peter Reutemann, and Ian H. Witten. 2009. "The WEKA Data Mining Software: An Update." *ACM SIGKDD Explorations Newsletter* 11 (1): 10–18. http://dx.doi.org/10.1145/1656274.1656278.

Hastie, Trevor, Robert Tibshirani, and Jerome H. Friedman. 2001. *The Elements of Statistical Learning: Data Mining, Inference, and Prediction*. Cham, Switzerland: Springer Science & Business Media. http://dx.doi.org/10.1007/BF02985802.

Khoo, Hui Chi. 2007. "Playing with Dynamics in the Music of Chopin." PhD thesis, Royal Holloway, University of London.

Kosta, Katerina, Oscar F. Bandtlow, and Elaine Chew. 2014. "Practical Implications of Dynamic Markings in the Score: Is Piano Always Piano?" In *Audio Engineering Society (AES) 53rd International Conference on Semantic Audio*, 26–29 January 2014, London, UK. http://www.aes.org/e-lib/browse.cfm?elib=17120.

Paderewski, Igancy Jan, Ludwik Bronarski, and Józef Turczyński. 2011. *Fryderyk Chopin, Complete works, Mazurkas*. 29th ed. Warszawa: Fryderyka Chopina, Polskie Wydawnictwo Muzyczne SA.

Quinlan, John Ross. 1986. "Induction of Decision Trees." *Machine Learning* 1 (1): 81–106. http://dx.doi.org/10.1023/A:1022643204877.

Repp, Bruno H. 1999. "A Microcosm of Musical Expression: II. Quantitative Analysis of Pianists' Dynamics in the Initial Measures of Chopin's Etude in E Major." *Journal of the Acoustical Society of America* 105 (3): 1972–1988. http://dx.doi.org/10.1121/1.426743.

Robnik-Sikonja, Marko, and Igor Kononenko. 1997. "An Adaptation of Relief for Attribute Estimation in Regression." In *Proceedings of the Fourteenth International Conference on Machine Learning (ICML '97)*, 8–12 July 1997, Nashville, TN, 296–304. San Francisco, CA: Morgan Kaufmann. http://dl.acm.org/citation.cfm?id = 657141.

Ros, María, Miguel Molina-Solana, Miguel Delgado, Waldo Fajardo, and Amparo Vila. 2016. "Transcribing Debussy's Syrinx Dynamics Through Linguistic Description: The *MUDELD* Algorithm." *Fuzzy Sets and Systems* 285, 199–216. http://dx.doi.org/10.1016/j.fss.2015.08.004.

Sapp, Craig. 2008. "Hybrid Numeric/Rank Similarity Metrics for Musical Performance Analysis." In *Proceedings of the 9th International Conference on Music Information Retrieval (ISMIR)*, 14–18 September 2008, Philadelphia, PA, 501–506.

Tae Hun, Kim, Satoru Fukayama, Takuya Nishimoto, and Shigeki Sagayama. 2013. "Statistical Approach to Automatic Expressive Rendition of Polyphonic Piano Music." In *Guide to Computing for Expressive Music Performance*. London: Springer.

Widmer, Gerhard, Sebastian Flossmann, and Maarten Grachten. 2009. "YQX plays Chopin." *AI Magazine* 30 (3): 35–48. http://www.aaai.org/ojs/index.php/aimagazine/article/view/2249.

Widmer, Gerhard, and Werner Goebl. 2004. "Computational Models of Expressive Music Performance: The State of the Art." *Journal of New Music Research* 33 (3): 203–216.

Data-based melody generation through multi-objective evolutionary computation

Pedro J. Ponce de León, José M. Iñesta, Jorge Calvo-Zaragoza, and David Rizo

Genetic-based composition algorithms are able to explore an immense space of possibilities, but the main difficulty has always been the implementation of the selection process. In this work, sets of melodies are utilized for training a machine learning approach to compute fitness, based on different metrics. The fitness of a candidate is provided by combining the metrics, but their values can range through different orders of magnitude and evolve in different ways, which makes it hard to combine these criteria. In order to solve this problem, a *multi-objective fitness approach* is proposed, in which the best individuals are those in the Pareto front of the multi-dimensional fitness space. *Melodic trees* are also proposed as a data structure for chromosomic representation of melodies and genetic operators are adapted to them. Some experiments have been carried out using a graphical interface prototype that allows one to explore the creative capabilities of the proposed system. An Online Supplement is provided and can be accessed at http://dx.doi.org/10.1080/17459737.2016.1188171, where the reader can find some technical details, information about the data used, generated melodies, and additional information about the developed prototype and its performance.

1. Introduction

Genetic algorithms (Holland 1975), or more generally, evolutionary techniques, are inspired by the biological evolution of living beings and natural selection. They are indeed optimization techniques in which a population of individuals, which represent different possible solutions, are subjected to processes that mimic natural crossovers and mutations, and a selection stage that decides which individuals are best fit to solve the problem. Those individuals are allowed to procreate a new generation of, supposedly, better individuals (solutions) until convergence. This basic idea covers a wide range of applications in engineering and machine learning (Goldberg 1989).

Music composition can be seen as an optimization process, in a way that the human composer implicitly searches in the space of all possible musical compositions for a composition that satisfies his or her own artistic criteria (Dostál 2013). Evolutionary-based algorithmic composition has always been well appreciated by researchers and composers due to their capability to stochastically explore an immense space of possibilities (Todd and Werner 1998). This is especially due to the mutation operator, which enables them to model something hardly computable like "inspiration" (something very vague that can be viewed as "finding something of artistic value by chance"). An interesting overview of the field can be found in the work of Miranda and Biles (2007).

In this paper, we aim to use machine learning techniques based on a training set of examples to implement an *automatic fitness function*. In the literature, this is considered to be one of the subtlest aspects of such systems (de Freitas, Guimarães, and Barbosa 2012). For that, we will try to assess different properties of melodies, some of them statistical, some music theoretical, combining them via the multi-objective optimization paradigm. This approach permits us to define a *fitness space*, rather than just a fitness function, where values that may be disperse can be put together in an elegant and easy way, in order to guide the composition process of the evolutionary system.

1.1. *Background: evolutionary computation in music composition*

Evolutionary computation in general, and genetic algorithms in particular, have been used in the arts over the years (Romero and Machado 2007), with growing interest. Biology-inspired computing techniques offer a metaphoric context for using nature as a source of inspiration, providing ideas and methods for finding other ways of solving problems and discovering new ways of creation.

In the case of algorithmic music composition, from the pioneering works of Xenakis (1992) or Hiller and Isaacson (1959), three main approaches have been used (Nierhaus 2008):

(a) *stochastic methods*, in which probability distribution functions are utilized to generate events that can be mapped to pitches, durations, or any other kind of psychoacoustic properties of music (like timbral or dynamical, for example);
(b) *deterministic methods*, which are based on rules (usually derived from musical methods or from mathematical theories) or sonification methods, by which mathematical functions (fractals, nonlinear dynamic systems, chaos theory, cellular automata, etc.) or signals and models coming from other areas of knowledge (physics, biology, geology, etc.) are used as data for being mapped to music parameters by using systematic transformations; and
(c) *machine learning (or data-driven) methods*, in which by using any kind of adaptive algorithm, usually based on examples of a music style or genre, the system makes its own model that is able to generate new data related to those previously used for training.

Of course, hybrid systems have also been proposed, providing very interesting results (Cope 1991), exploiting the best qualities of each approach, at the cost of increasing the system's design complexity.

Genetic algorithms (GAs) belong to the third category of methods, where the composition process is guided somehow, aiming towards a particular kind of music. The main constituent parts of a GA are a representation for chromosomes (the candidate solutions), an initial population of chromosomes, a set of operators to generate new candidates, an evaluation function, and a selection method.

The most usual representation of a musical chromosome is a vector of numbers (often in binary code), representing pitch and duration information. Some authors have explored other approaches, like analytic functions (Laine and Kuuskankare 1994), languages like `abc` (Oliwa 2008), or motives represented by patterns (Liu and Ting 2015). More often, trees have also been used (Hofmann 2015; Komatsu et al. 2010; Phon-Amnuaisuk, Law, and Ho 2007), but always in the context of a genetic programming approach, where the trees are representing a string in a language, formally represented by a syntax, so they are actually *parsing trees*. In the present work, we introduce the use of *melodic trees* for this task, as a structured representation of a monophonic series of notes. Although the trees in the GA implementation are defined by a grammar, they represent structural melodic trees. We will need to define the proper genetic operators, adapted for this representation.

1.1.1. *Style-oriented approaches*

Music style is a vague concept that can be interpreted in a number of ways, all of them subjective. Some authors consider the music style as the set of music compositions that provide a given emotion (aggressive, calm, romantic, etc.) while others focus on music genre as the set of authors or pieces sharing some common characteristics.

A style-guided genetic composition scheme can be achieved by using machine learning algorithms and a corpus of training data from a music genre, a given author, or even particular tastes, to implement fitness functions or any kind of guidance in the composition process (Alfonseca, Cebrián, and Ortega 2007; de Freitas, Guimarães, and Barbosa 2012; Weale and Seitzer 2003). Thus, the same overall scheme can be used to generate very different melodies just changing the training data. Such a corpus must be tagged before. This is usually done by a human, so a subjective factor is introduced.

1.1.2. *Human supervised fitness functions*

One of the key aspects of every evolutionary system is the design and implementation of the fitness functions. This is particularly tricky when the individuals that are being evaluated are artworks, although there are systems like GenDash (Waschka 2007), which is a GA without fitness evaluation, where the individuals are selected at random to the next generation, avoiding what is considered by the author as the *bottleneck* of evolutionary systems. There is not an ultimate evaluation system; even the impressions of some persons listening to the same music can be different and, therefore, the results of automatic composition are subjective.

The aid of human skills for that permits to model the musician's taste and even web-based human assessment of the individuals have been described in the literature (Ayesh and Hugill 2005; Fu et al. 2006; Koga and Fukumoto 2014; Özcan and Erçal 2007; Putnam 1996; Tokui and Iba 2000; Zhu, Wang, and Wang 2008). Sometimes, the user is needed because the system is oriented to user interaction, like in the case of the GenJam system (Biles 1994), but sometimes the user is needed because the system is expected to accelerate the optimization by the user actions (Koga and Fukumoto 2014), like evaluating, removing, or even changing the melodies generated at a given generation. In Zhu, Wang, and Wang (2008), the system is focused on capturing listener feelings, like happiness and sadness, although the user works in cooperation with a rule system to calculate the fitness values.

In most of these cases, the populations are reduced to a very small number of individuals. It is obvious that evaluating hundreds or thousands of music works is not feasible, even when many users are involved, like in Ayesh and Hugill (2005). Anyway, when human critics are used, these evolutionary systems can produce pleasing and sometimes surprising music, but usually after many tiresome generations of feedback (Todd and Miranda 2003).

1.1.3. *Autonomous fitness functions*

Given the problems and limitations of human-supervised assessment, there are many works in the literature that deal with the problem of designing fully automatic evaluation functions that can operate without human intervention. This idea of automatic evaluation of music (and artworks in general) has received the name of a virtual "critic" (Machado et al. 2003).

Some works (Wiggins et al. 1998) use music-theoretical knowledge as a fitness function, like Moroni et al. (1994), who use a fitness criterion that takes into account melodic fitness, harmonic fitness, and voice range fitness. Other previous works try to induce musical structure from a corpus to get new melodies (Cope 1991). An extension of this procedure was done by Spector and Alpern (1995), who applied a hybrid rule-based and neural network critic trained with a corpus of well-known works, and Baluja, Pomerleau, and Jochem (1994) who also used a similar combination of neural and genetic techniques.

Language models, like *n*-gram models (Herremans, Sörensen, and Conklin 2014; Lo and Lucas 2007) have been used as trainable music critics to impose constraints on the space of pitches and intervals that can be explored by the genetic algorithm. This idea will also be explored in the present work.

Neural techniques have been widely used to implement fitness functions. For example, in Sheikholharam and Teshnehlab (2008) recurrent networks are used both for generating pitch sequences and for evaluation, in particular to cope with the problem of long-term melodic relations, which is impossible to capture using language models, like *n*-grams. In Göksu, Pigg, and Dixit (2005) multilayer perceptrons are trained to recognize music of a given genre. Also, "self-organized maps" were used in Phon-Amnuaisuk, Law, and Ho (2007), exploiting their ability to map music representations into feature visualization spaces.

1.1.4. *Multiple fitness values*

A valuable approach that is currently gaining ground is the use of different fitness functions, each of them specialized in a particular aspect of evaluation, providing different values that are combined to get a better combined fitness. The key point here is to find an appropriate way of combining those values.

For example, in Ayesh and Hugill (2005) the selection process is guided by the responses of the users within an interactive process. The individuals are rated by 3 different factors that are combined by simple summation into a single fitness value. In the context of automatic fitness, in Özcan and Erçal (2007) multiple musical aspects are evaluated: 10 fitness functions are computed and then combined by using a weighted sum of their respective evaluations. Hofmann (2015) introduces the concept of multi-objective evaluation, which is central in the present work, but what that author actually does is look for the optimum of each of the 11 statistical and 6 structural fitness function modules utilized. Each function is considered as a vector dimension and the system tries to optimize a weighted distance from each individual to the optimum vector.

A proper strategy for combining the different values provided by a number of fitness evaluation functions is needed. Summing them is not appropriate since they can differ even in orders of magnitude, and therefore it is not straightforward to find a suitable way to merge them into a single criterion.

It is at this point where a multi-objective optimization (MOO) approach to establish an ordering relation among individuals is useful, taking into account the whole set of criteria simultaneously. Although the combination of multi-objective optimization with genetic algorithms has been widely studied (Purshouse et al. 2013), the proposed application for music composition is innovative. In the context of music recognition, a previous work (Vatolkin 2013) used MOO for prediction of high-level music categories, such as genres, styles, or personal preferences. That author measured the impact of evolutionary multi-objective audio feature selection on the classification performance, leading to a significant reduction in the classification error.

1.2. *The present approach*

In the present approach, two main novelties are introduced. The first novelty is the use of a tree data structure (Rizo 2010) for coding the individuals. Although, as discussed above, trees have been used before in evolutive composition, they have always been used for coding a programming language string as a formal grammar, so they are actually a tool for implementing genetic programming methods. Standard genetic programming mutation and crossover operators are applied on this data structure. In our case, a *tree* is a representation of a melody that will compete for survival. The target will be melodies of a fixed length in bars (eight bars by default). A proof of concept of this approach was already presented in (Espí et al. 2007).

The second novelty of the present paper is the combination of different fitness functions by means of a multi-objective optimization method that is able to combine multiple and diverse values in order to rank the individuals in a single fitness space. This way, the individuals are not ranked by their fitness values, but instead by their position in the fitness space. This way an optimal locus is defined by some of the individuals that will be ranked as the best adapted. After them, other individuals will be in the ranking by selecting the best after removing the former ones from the fitness analysis. And so on.

The fitness dimensions in this space will be provided by example-based machine learning models, some of them based on statistical properties of the melodies, others based on statistical language models, and others designed on top of musicology rules, trained from the melodic examples contained in a provided training set.

2. Methods and tools

2.1. *Melody tree representation*

There are several ways of coding a melody (Selfridge-Field 1997). In this work, we use the tree encoding proposed by Rizo (2010), extended with the addition of levels for sections and measures. That model is based on the representation of the rhythm structure as a tree. Each bar is represented as a sub-tree, depending on its meter. All these sub-trees are linked together in a common tree, with a binary or ternary arity, depending on the kind of meter. The level of a node determines its duration: the root represents the duration of the whole melody, the nodes of the next level represent a division of the upper node that, together, sum up its whole duration.

Pitch codes are found in the leaves of the tree. Any kind of absolute or relative pitch could be used. For this particular application, pitch classes, together with rest and continuation symbols, have been used: $p_i \in \{C, C\sharp/D\flat, D, \ldots, B\} \cup \{rest, continuation\}$. Some pitch classes from the whole chromatic scale can be selected, if required.

Figure 1. A short melody example and its corresponding tree representation as a section of what can be a longer melody. The labels in the leaves correspond to pitch classes, including "r" for rests, and " – " for continuations. "S" denotes a *section*, "M" is a *measure* node, and "∧" represents every inner node (beat or sub-beat) that splits.

These two extra symbols are used for coding rests and the continuation of a note beyond its natural duration, like in the case of syncopations, ties or dotted notes (see Figure 1 for an illustration). The left to right ordering of the leaves encodes the onset times of the notes in the melody.

Without loss of generality, we shall restrict this study to binary trees, representing binary subdivisions in measures and beats, diatonic pitches from the major scale, and maximum depth for the trees corresponding to a sixteenth note.

Another interesting feature of this coding is that it is very easy to change the structure of the melody tree representation in the genetic algorithm, by changing the underlying grammar in a configuration file. Currently, the structure that is being used is as follows:

```
Melody := Section Section
Section := Measure Measure Measure Measure
Measure := (Beat | Symbol) (Beat | Symbol) (Beat | Symbol)
           (Beat | Symbol)
Beat := (SubBeat | Symbol) (SubBeat | Symbol)
SubBeat := Symbol Symbol
Symbol := Pitch | Rest | Continuation
Pitch := C | D | E | F | G | A | B
Rest := r
Continuation := _
```

but it can be changed by specifying another structure for the system.

Note that one of the main advantages of this representation is that it is not possible to generate invalid representations when crossover operations mix the chromosomes of the different individuals, which is very useful in the evolutionary paradigm.

2.2. *Evolutionary algorithms with trees*

Each individual of our population will be a 4/4 meter, 8-bars long, monophonic melody coded as a binary tree, as described above.

The size of the population is a free parameter of the system. The population is initialized at random following the tree generator procedure proposed by Koza (1992), and the evolutionary search loops. Hypothetically, better compositions are obtained through the evolutionary crossover and mutation operators. Individuals are evaluated and selected so that only those who encode the best solutions, according to their *fitness*, can survive for the next generation.

The crossover operator creates two individuals (children) from two existing ones of the previous generation (parents). The idea is that the children contain features of their parents, but represent new melodies. For this, one non-leaf node is randomly selected in each of the parents. The whole sub-tree hanging from the selected node is interchanged by that in the other parent tree. Due to the representation utilized, the bar length is preserved. Measure nodes can also be interchanged by crossover (see Figure 2).

The mutation operator is applied on each individual separately. It consists of randomly changing its chromosome. Each node of the tree has a small probability of being modified. In that case, the mutation operator generates a sub-tree whose root is compatible with the kind of the selected node, according to the grammar structure displayed above in Section 2.1.

By applying these operators, an offspring is created from a population of size N. It is done as follows: 90% of the time, a series of two tournaments of seven individuals, randomly chosen, are carried out. The winners of each are used as parents for a crossover process, breeding two new individuals. This procedure is repeated until generating N individuals. The new individuals are

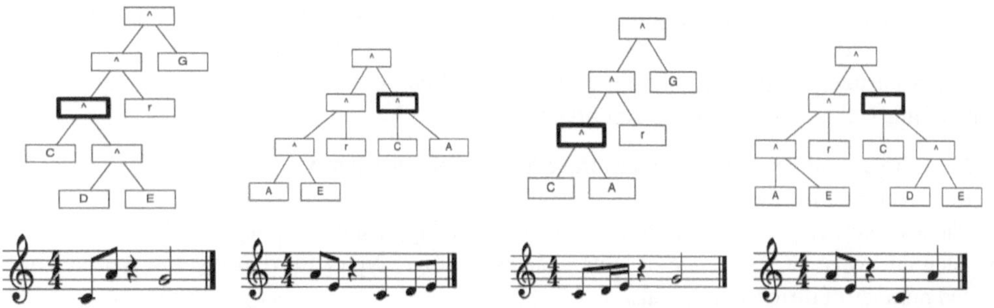

Figure 2. Illustration of the crossover operator applied on 1-bar trees in 4/4, and the effect on the melodies they represent. The two parents are those displayed on the left, and on the right the two children are shown. A node is selected for both parents (thick node borders) and the sub-trees are interchanged.

assessed and ranked together with their parents. In the remaining 10% of the time, the population is simply cloned. Note that, whatever the case, the size of the original population is doubled.

From the new individuals generated by the crossover, the mutation pipeline is computed. It consists in performing tournaments of 7 individuals for which the winner is mutated. At the end, each mutated individual will be evaluated by the set of selected fitness metrics, as will be described in the next sections.

At the end, only the best N individuals among the parents and the offspring are kept to maintain the size of the population. The survivals are considered for the next generation. The process starts over again, until reaching a maximum number of iterations, which can be set by the user.

2.3. *Automated feature extraction and fitness*

One of the goals of this work is to study how to use different machine learning techniques to rate evolutionary generated music according to the parameters tuned from melodies in a set, which supposedly share some common properties, like genre, style, mood, etc. For that, the evolving melodies have to be described using the features that are the input to those assessment systems.

In this regard, we have considered three families of fitness evaluations:

(a) global statistical descriptors that embed the melodic tree in a vector space representing the whole melody as a point. This way, similarities can be computed as distances in such a space;
(b) local n-gram probabilistic models for assessing how likely it is that notes could be used together in a short-term context; and
(c) music theory-based melodic evaluations, in order to introduce domain-specific knowledge in the fitness evaluation process.

These three families comprise a total of 12 separate fitness functions. By using these functions, we create a *fitness space* in which the multi-objective optimization algorithm is used to select the best individuals. The final user may decide to use all or a part of those functions. For that, a graphical interface has been designed for experimentation that will be introduced in the experiments and results section.

2.3.1. *Global statistical evaluations*

The first model is the *global shallow description scheme* (Ponce de León and Iñesta 2007). In this case, an individual melody is represented as a vector of statistical descriptors that are related to melodic, harmonic, and rhythmic properties of the melody, by analysing how pitches, silences, durations, inter-onset intervals, pitch intervals, diatonic notes, and syncopations are distributed in the melody (the reader is referred to the paper cited above for details).

In this particular implementation we have selected the following descriptors, grouped by the kind of musical property they are describing.

- Notes: total number of notes and average number of notes per beat.
- Rests: total number of silences.
- Pitches: average pitch relative to the lowest one and typical deviation.
- Pitch intervals: largest, range of interval sizes, average, typical deviation, most repeated one, and number of different intervals.
- Durations: occupation rate (sum of note duration versus total duration), duration range (from shortest to longest), average duration relative to the shortest note, and typical deviation.
- Inter-onset intervals: range (from shortest to longest), average relative to the shortest one, and typical deviation.
- Rest durations: range (from shortest to longest), average relative to the shortest rest, and typical deviation.
- Syncopations: total number of syncopated notes and rate (syncopated versus total number of notes).

These descriptors create a vector $\mathbf{x} \in \mathbb{R}^{23}$. This way, the melodies $\mathbf{x}_i$ in the training set $\mathcal{X}$ are represented as a cloud of points that are our target style. Under the hypothesis that new melodies of that style should be close to those already existing, the distance from the vector $\mathbf{m}$ representing each evolving melody to the cloud is given as a measure of style for it.

The global centroid of the cloud, $\mathbf{c} = \frac{1}{|\mathcal{X}|}\sum_i \mathbf{x}_i$, can be a target for that matter, but if all the population $\mathbf{m}_j$ tries to minimize the distances $d_c = d(\mathbf{m}_j, \mathbf{c})$, this may not favor diversity. For that, the distance to be minimized should be more local, and we have considered the distance to a neighborhood of size k of each individual, in such a way that a local centroid is computed as $\mathbf{c}_k = \frac{1}{k}\sum_{i=1}^{k} \mathbf{x}_i$ and the distance to be minimized is $d_k = d(\mathbf{m}, \mathbf{c}_k)$. The problem with this approach is that the point $\mathbf{m}$ may be far from the global centroid, which is supposed to be a good reference for the style of music being represented. Taking into account the pros and cons of the two approaches, both distances, d_c and d_k, have been used for computing fitness functions to be minimized.

In any case, the features computed are very different, so in order to cope with the variety of ranges and dispersions they may show, the Mahalanobis distance has been used:

$$d_c^2(\mathbf{m}, \mathbf{c}) = (\mathbf{m} - \mathbf{c})^{\mathrm{T}} \Sigma^{-1} (\mathbf{m} - \mathbf{c})$$

$$d_k^2(\mathbf{m}, \mathbf{c}_k) = (\mathbf{m} - \mathbf{c}_k)^{\mathrm{T}} \Sigma^{-1} (\mathbf{m} - \mathbf{c}_k),$$

where Σ is the covariance matrix in $\mathbb{R}^{23\times 23}$ computed from $\mathcal{X}$.

Once the proposed distances are computed as $d^2 = \{d_c^2, d_k^2\}$, the fitness values $F_{\mathrm{global},d}$ are normalized by using

$$F_{\mathrm{global},d} = 1 - \frac{1}{1+d^2} = \frac{d^2}{1+d^2}.$$

This value is what the genetic algorithm will try to minimize.

2.3.2. *Local musical n-gram evaluations*

The fitness functions $F_{\mathrm{global},d}$ provide a global statistical context for the possible melodies, driven by the training examples represented in the feature vector space. Nevertheless, there are still a huge number of melodies that could give the same global statistics, so a more local, restrictive, view is needed. This way, more restrictive constraints are imposed on adjacent notes. We could say that proximity to the cloud representing the training set provides a necessary condition but not a sufficient condition for a melody to sound style-like.

An *n-gram model* is a statistical language modeling technique widely used in natural language processing, which works under the Markov assumption that the probability of a symbol depends only on the probability of a short-term history of previous symbols. Each parameter is the probability of a symbol s_i to appear after seeing the sequence of $n-1$ symbols $s_{i,n-1} := s_{i-n+1}s_{i-n+2}\cdots s_{i-1}$. This probability is estimated from the relative frequency of the string $s_{i,n} = s_{i,n-1}s_i$ in the training set, as

$$P(s_i|s_{i,n-1}) = \frac{\mathcal{N}(s_{i,n-1}s_i)}{\sum_{s\in\mathcal{A}} \mathcal{N}(s_{i,n-1}s)} .$$

Here $\mathcal{A}$ is the alphabet of possible symbols from which s_i can take values, and $\mathcal{N}$ is a counting function. One of the problems with this approach is that, if a sequence of n symbols does not appear in the training set, its probability is zero, and therefore the probability of any sequence in which it may appear will also be zero. To solve this problem, the probability distribution is smoothed by using the Kneser–Ney method (Kneser and Ney 1995), reallocating some probability mass from the 3-grams or 4-grams utilized, to simpler unigram models.

Once we have the n-gram model probabilities inferred, we can assign a probability to a new sequence S of $|S|$ symbols by computing

$$P(S) = P_{n-1}(s_1\cdots s_{n-1}) \prod_{i=n}^{|S|} P(s_i|s_{i,n-1}) ,$$

where $P_{n-1}(s_1 \cdots s_{n-1})$ denotes the probability of a string beginning with the $n-1$ symbols that cannot be computed with the n-gram model, and is estimated by looking for how often the strings in the training set begin with that sub-string, $\mathcal{N}(s_1 \cdots s_{n-1})/\mathcal{N}(S)$.

The length of the string coding the melody, $|S|$, is an issue in this case, because the lower the number of n-grams, the higher the probability will be, and vice versa. To cope with that problem, the length of the melody is also modeled as a Poisson distribution

$$p(|S|, \lambda) = \frac{e^{-\lambda}\lambda^{|S|}}{|S|!},$$

where the expected value for the length, λ, is estimated from the melodies in the training set, $\mathcal{X}$. So eventually, the probability of a sequence S will be

$$P(S) = p(|S|, \lambda)P_{n-1}(s_1\cdots s_{n-1}) \prod_{i=n}^{|S|} P(s_i|s_{i,n-1}).$$

For using these models in our case, we construct strings of symbols (*n-words*) (Doraisamy and Rüger 2003) from n consecutive notes in the tree representation as a string of $n-1$ intervals coded by symbols in an alphabet $\mathcal{A}_P$ and $n-1$ inter-onset ratios (IOR) coded by symbols in an alphabet $\mathcal{A}_D$ (see details in the cited paper). This way, for example, a series of $n=3$ notes is coded in a *coupled* representation of relative pitches and durations by 4 characters in a '3-word'. There is also the possibility of *de-coupling* the representation by computing separately 3-words with the 2 intervals and with the 2 inter-onset ratios. This may be useful if the data available are limited, because for the coupled representation, the cardinality of the alphabet is $|\mathcal{A}| = (|\mathcal{A}_P| \times |\mathcal{A}_D|)^{n-1}$, while for the decoupled representations the cardinality of the alphabet is just $|\mathcal{A}_P|^{n-1}$ or $|\mathcal{A}_D|^{n-1}$. These lower cardinalities make the parameters easier to learn from the training melodies.

All possible n-words are extracted from an individual melody, except those containing a silence lasting four or more beats, which are ignored. From the melody coded as a series of n-words, the *n-gram sequence*, S, of n-words is analysed and its probability $P(S)$ is computed, and the fitness function will be

$$F_{\text{local},n\text{-gram}} = -\log P(S)$$

for $n = \{3, 4\}$, and for the coupled (intervals and durations together) and decoupled (either for intervals or IORs) versions of the n-word coding.

2.3.3. *Local bag-of-notes evaluations*

The other approach to style modeling is similar to that above in terms of coding n notes into a series of characters by using an n-word, but focuses on how probable a string based on the probability of appearance of substrings of n-words is, rather than in the order of how they appear, like in the former case.

The same n-word based representation described above for a melody (only the coupled version for $n=3$) is now evaluated probabilistically by a multinomial distribution model that takes into account n-word frequencies in the string. In this model, each melody is encoded as a vector $\mathbf{s} \in \mathbb{N}^{|\mathcal{V}|}$, where $\mathcal{V}$ is the vocabulary made of the most frequent n-words found in the training set, and each component s_t represents the number of occurrences of the n-word w_t in the melody.

The same problem with the string coding length $|S|$ as in the former case is present here, so shorter notes would imply fewer products, and therefore always higher probabilities, so a new Poisson distribution is estimated from the training data, $p(|S|, \lambda)$, by computing the average number of n-words found in 8-bar melody segments.

The probability that the whole series of n-words in the melody has been generated by the multinomial distribution found for the melodies in the training set $\mathcal{X}$ is

$$P(\mathbf{s}|\mathcal{X}) = p(|S|, \lambda)|S|! \prod_{t=1}^{|\mathcal{V}|} \frac{P(w_t|\mathcal{X})^{s_t}}{s_t!}$$

and the conditional word probabilities are estimated as

$$P(w_t|\mathcal{X}) = \frac{1 + \mathcal{N}_t}{|\mathcal{V}| + \sum_{k=1}^{|\mathcal{V}|} \mathcal{N}_k},$$

where $\mathcal{N}_t$ is the sum of occurrences of word w_t in the melodies of the training set.

Like in the former case, once the $P(\mathbf{s}|\mathcal{X})$ is computed, the fitness function will be

$$F_{\text{local,multi}} = -\log P(\mathbf{s}|\mathcal{X}) .$$

2.3.4. *Melodicity fitness evaluations*

A complementary approach to the fitness evaluations described in the sections above is based on elements of music theory, which permits one to implement knowledge specific to the application. Music theory provides a general conceptual framework that can be adapted to the data in a training set by learning parameter values and distributions.

The kind of evaluations we are going to perform are two-fold: on one side, a melodic analysis that tags the notes in the melody as harmonic or non-harmonic (with sub-classes) and then a language model inferred from the training data is applied over the tags, and on the other side, a segmentation algorithm is applied and the number of segments is compared to the number of segments found in the training set for melodies of the same length.

2.3.5. *Melodic analysis*

Melodic analysis determines the importance and role of each note in a particular harmonic context. Thus, a note is classified as a harmonic tone ("H"), when it belongs to the underlying chord, and as a non-harmonic tone otherwise, in which case it should be further assigned to a category, such as *passing tone* ("P"), *neighbor tone* ("N"), *suspension* ("S"), or *appoggiatura* ("A"). There are more possible categories (see Willingham [2013]), but the ones considered here are those based on the stability of the beat and the intervals before and after the analysed note. There are other kinds of non-harmonic notes, but the features required for performing the analysis need harmonic information, like chords, that are not available for a monodic melody (see Rizo, Illescas, and Iñesta [2015] for details).

A n-gram model is constructed from the series of melodic tags obtained when the fragments of 8-bar melodies in the training set are analysed. The same approach as in Section 2.3.2 is applied to the 5 tags, listed above $\mathcal{A}_A = \{\text{"H"},\text{"P"},\text{"N"},\text{"S"},\text{"A"}\}$ and the probability $P(A)$ of the analysis sequence $A \in \mathcal{A}_A{}^\star$ of an individual is computed. From it,

$$F_{\text{melodic},n} = -\log(P(A))$$

is established as a fitness value to be minimized for $n \in \{3, 4\}$.

2.3.6. *Melodic segmentation*

The *local boundary detection model (LBDM)* (Cambouropoulos 2001) is a simple and well-known algorithm that permits one to partition a melody into segments using the sizes of intervals, the length of notes, and the length of silences. These three measures are weighted by normalized coefficients, w_i, that tune the algorithm's behavior. We have heuristically set them to $w_1 = 0.6$, $w_2 = 0.2$, and $w_4 = 0.2$, respectively, in order to give more importance to intervallic relations. The threshold over the boundary strength for local maxima detection was $\theta = 0.3$.

An excessive number of segments, σ, denotes a lack of coherence in the melody, while a very low number of segments is an indication of a dull, flat melody. In any case, a good number of segments depends on both the total length of the melody and the style of music. In order to adapt this to data, we have made a statistical study on how many segments per fragment are in the training set, considering melodies of the same length as those generated by the genetic algorithm (8 bars in this case). A new Poisson distribution $P(\sigma) = p(\sigma, \lambda)$ was estimated from the number of segments, and the fitness to be minimized will be $1 - P(\sigma)$, in order to favor a number of segments close to λ.

2.4. ***Multi-objective optimization***

Without loss of generality, we are going to consider that optimizing a function refers to its minimization. Therefore, a typical optimization problem can be defined as finding an $\hat{x}$ such that

$$\hat{x} = \arg\min_{x \in \mathcal{S}} f(x)$$

in which $\mathcal{S}$ stands for the set of solutions that fulfill the implicit constraints of the problem.

Quite often, applications require us to have multiple functions or objectives to be optimized at the same time. In fact, a total number of $F = 12$ fitness functions have been proposed to implement the evaluation of a music composition. Then, the optimization problem is reformulated as

$$\hat{x} = \arg\min_{x \in \mathcal{S}} \{f_1(x), f_2(x), \ldots, f_F(x)\} \ .$$

In this case, the definition of an optimal solution cannot be taken directly from the scalar concept of the single-objective optimization. When multiple functions have to be optimized simultaneously, optimality is defined through the concept of *dominance*.

Definition 2.1 A point x_1 in $\mathcal{S}$ is said to *dominate* another point x_2 in $\mathcal{S}$ if, and only if, both of the following conditions hold.

(1) The point x_1 is better than or equal to x_2 in every single objective function.
(2) The point x_1 is strictly better than x_2 in at least one of the objective functions.

The set of non-dominated solutions, those for which none dominates them, is referred to as the *Pareto front*. The elements within this set have the property of being *Pareto optimal*, which means it is impossible to make an improvement in one criterion without making at least another one worse.

In this paper we are interested in developing an evolutionary algorithm that is able to optimize several functions simultaneously. Hence, the solution space is not explored exhaustively but new points are obtained through iterative generations of an evolutionary search. To make this process generate solutions that converge to the optimal values, it is necessary to establish an order relationship among individuals in the population that allows selecting those most promising in relation to the concept of Pareto optimality. To this end, this work follows a multi-objective evolutionary scheme.

A number of ways of addressing this problem have been proposed over the last years. In a first batch of algorithms (Horn et al. 1994; Schaffer 1985; Srinivas and Deb 1994), non-dominated individuals of the population were kept through the evolutionary generations. Although these strategies were able to find Pareto-optimal solutions, the set of surviving individuals tended to concentrate on a small portion of the target space, in which only one of the functions was actually optimized. That is why a second group of algorithms arose (Deb et al. 2002; Zitzler, Laumanns, and Thiele 2001), which also promoted the diversity of solutions throughout the Pareto front. An example of this algorithm is the Non-dominated Sorting Genetic Algorithm II, which will be used in this work. The next section briefly describes the operation of this algorithm given its interest in the present paper.

2.4.1. *The Non-dominated Sorting Genetic Algorithm II*

The *Non-dominated Sorting Genetic Algorithm* II (NSGA-II) is an efficient multi-objective optimization evolutionary scheme proposed by Deb et al. (2002). Although it was developed for genetic algorithms, NSGA-II can run on any evolutionary search that contains evaluation and selection processes, as it focuses on establishing a relationship of order or priority among individuals in the population according to their multi-valued fitness.

The general operation of the algorithm is to divide the individual into fronts of non-dominance. In the first front, the algorithm finds those individuals that are non-dominated (Pareto front); in the second front, it finds those individuals that would form a new optimal front in the absence of the individuals of the first one; and so on. These fronts are iteratively created until every single individual in the population is assigned to one.

Once this process is over, a new relationship is established, but only among individuals belonging to the same rank. For this purpose, NSGA-II defines a *crowding distance* function to estimate the diversity that brings each of the individuals within that front. The idea is to favor those individuals that provide a higher variety of solutions.

Formally speaking, NSGA-II defines a partial relation among the individuals of the population. Let $P = \{p_1, p_2, \ldots, p_{|P|}\}$ be a population. We denote by $r : P \to \mathbb{N}$ the front assigned to an element of P and by $c : P \to \mathbb{R}^*$ the crowding distance function. NSGA-II establishes a partial order ($\succ$) among the elements of P such that

$$p_i \succ p_j \Leftrightarrow (r(p_i) < r(p_j)) \vee (r(p_i) = r(p_j) \wedge c(p_i) > c(p_j)). \tag{1}$$

Therefore, when the evolutionary algorithm reaches the selection step, individuals of the population, as well as the offspring generated from the crossover and mutation operators, are divided into non-dominance fronts $(F_1, F_2, \ldots)$. Within each rank, individuals are ordered according to the crowding function. Eventually, the selection operator takes the individuals in descendant order until reaching the criteria established to maintain the population size for the next generation.

3. Experiments and results

It is very difficult to show and evaluate the performance of a system that is designed to generate artworks. We can show that there are signs of convergence: the fitness functions that have to be minimized are actually decreasing their values, the Pareto front is getting closer to **0** in the fitness space, etc. Next, some of these indicators will be presented. But, in addition, it is very important to be able to experiment with the different fitness functions and parameters, especially in the case of techniques like evolutionary computing that have so many free parameters.

For that, a graphical interface prototype implemented in Java has been developed (see the Online Supplement for details). The genetic algorithms were programmed using the Java-based Evolutionary Computation Research System (ECJ) library.[1] All the mathematical calculations were programmed from scratch, except for the n-gram models, for which the BerkeleyLM library (Pauls and Klein 2011) was used.

The prototype permits one to select a training dataset in the form of single-track monophonic MIDI files contained in a folder. Some information about the datasets used for testing our system is provided in the Online Supplement. There is also the possibility of changing the configuration parameters for the genetic algorithm by a configuration file in text format. The prototype is available for download from the URL provided.[2]

Concerning the running time, it takes about one minute on an iMac with a 2.7 GHz Intel Core i5 processor to run a 1000-generations evolution of a population of 100 individuals, using the 12 different fitness functions that can be utilized. If only four functions are used (one global, one local n -gram, one local multinomial probability, and one melodic segmentation) the running time is reduced to 24 seconds for the same population and number of generations.

Figure 3 displays how the Pareto front frontier evolves over the generations for two pairs of fitness functions. Note how it gets closer to the origin (the algorithm is minimizing the functions), with longer jumps at the beginning and then stabilizing in the last generations. It is not possible to represent the actual Pareto front in the 12-dimensional space, but just a selection of two dimensions, like those in Figure 3. There are more plots of the fitness evolution for pairs of functions in the Online Supplement.

[1] https://cs.gmu.edu/ ~ eclab/projects/ecj/

[2] http://grfia.dlsi.ua.es/gen.php?id = software

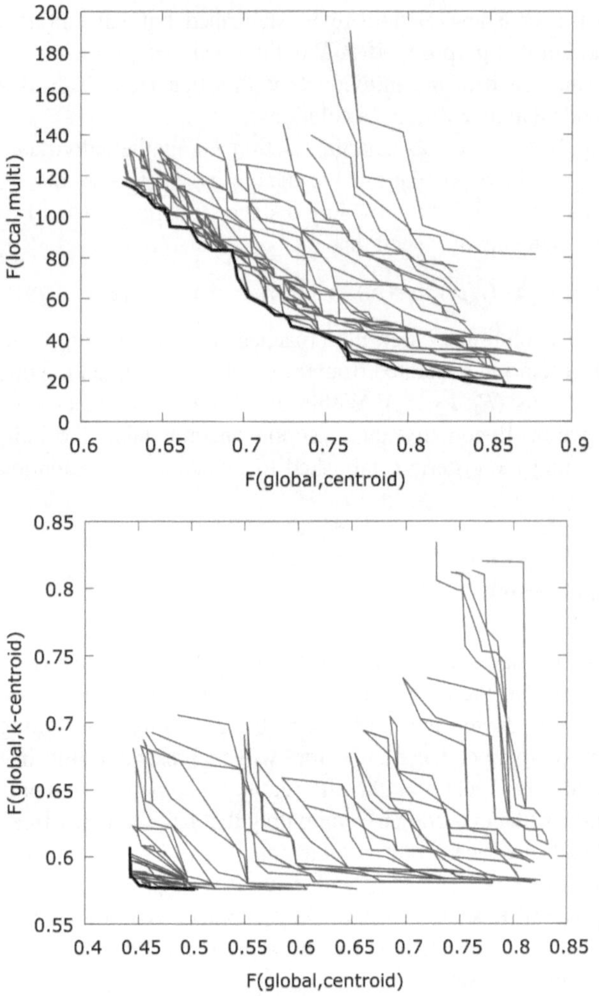

Figure 3. Evolution of the Pareto front over the generations. It is depicted for two pairs of fitness functions. Upper: $F_{\text{local,multi}}$ and F_{global,d_c}. Lower: F_{global,d_c} and F_{global,d_k}. It starts closer to the upper-right corner and it moves towards $(0,0)$. The Pareto front at the end of the iterations is highlighted.

3.1. *Single fitness function analysis*

In this section, salient features of phenotypes obtained by the application of single objective functions to the population are described. The fitness functions were trained on a dataset made up of MIDI files of popular music melodies from three subgenres: blues, pop, and celtic music. The system is designed to produce 8-bar melodies in the 4/4 meter.

The melodic phenotypes are described qualitatively in terms of their:

structure: presence of repeated melodic patterns,
segmentation: presence of phrase boundaries within the melodic sequence,
pitch: presence of implicit tonality, pitch variety, size of intervals,
note durations: most frequent note durations, or IOI, and
convergence: has the system converged to local minima? Is phenotype variety preserved across generations?

The analysis is done manually, in search of strong evidence for the melodic features described above in the best individuals from the final population of several runs.

Structure. The n-gram based local evaluators produced melodies that often had repeated patterns. This was also the case with global statistical evaluators, although the repeated patterns were of shorter length, on average (for example, half a measure).

Segmentation. Both segmentation and melodic analysis evaluators produced phenotypes with clear phrase boundaries. Global statistical evaluators, on the other hand, seldom generated phrase boundaries. There was no clear evidence of those boundaries in phenotypes produced by other evaluators.

Pitch. Segmentation and local evaluators based on n-grams using pitch information often produced individuals with an excess of unisons. However, when duration-based n-grams were used, there was no such problem. Local bag-of-notes models produced degenerated melodies containing often a single n-gram or no notes at all. Melodies generated by melodic analysis evaluators often had a sense of tonality, with evidence of pitch variety, but rarely with the presence of intervals wider than a fifth. Global statistical evaluators produced a fair variety of pitches in melodies.

Note durations. Higher variety in terms of note durations was found when using global statistical evaluators, while the others tended to produce rhythmically monotone melodies. Local n-gram models based on duration, for example, mostly produce melodies containing quarter and half note durations. Segmentation fitness produced many more of sixteenth notes compared to other evaluators, while the remaining criteria produced mainly eighth or quarter notes.

Convergence. Figure 4 shows two examples of fitness convergence from different evaluator functions. Local bag-of-notes, melodic analysis, local n-grams based on pitch and duration, and

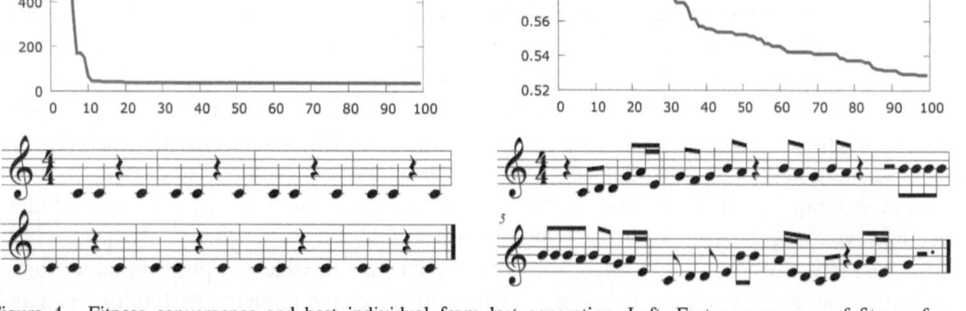

Figure 4. Fitness convergence and best individual from last generation. Left: Fast convergence of fitness for a $F_{\text{local},4\text{-gram}}$ evaluator function. Right: Smooth convergence for a F_{global,d_c}. Both evaluators were trained on a dataset of popular music. See the Online Supplement for details of the dataset.

local n-grams based on pitch-only evaluators exhibited fast convergence (less than ten generations, on average) to local minima when tested in isolation. Local n-grams based on evaluators converged slowly, while global statistical functions converged at slow rates, an indication that they are less prone to be trapped too soon on local minima. When used as single fitness evaluators, all functions led the evolutionary process to converge to local minima most of the time, and to produce a rather homogeneous population.

3.2. *Combined fitness function analysis*

When considering pairs of fitness functions, the performance outcome improved when using them in isolation, but they still fell into local minima. For example, when one of the evaluators used trigrams based on pitch, flat melodies still appeared, with many successive unisons.

By combining all families of evaluators, the results were closer to what one might expect. Even so, melodies clearly affected by features inherent in certain fitness functions appeared, but these were often compensated by other evaluators. For example, the result of using an evaluator that tends to produce flat melodies in terms of pitch or monotonous durations of the notes may be balanced by other evaluators that favor the dispersion of pitches and durations. Those evaluators that contribute to the repetition of patterns and tend to produce excessive repetition can help in producing some structure in the melody. On the other hand, their harmful effects are mitigated by functions that favor melodic diversity. In any case, this operation is heavily modulated by the data used as the basis for the training of evaluators.

4. Discussion and conclusions

Genetic algorithms present interesting features for music creation. Systems like those presented here can be useful for the generation of interesting musical ideas that can be developed or refined by human composers.

The main issue with this kind of technique has always been how to evaluate the fitness of maybe hundreds of candidate individuals along thousands of generations, which makes the participation of human assessments in this kind of system unfeasible in practice, unless the size of the population is reduced to a few individuals. Machine learning and pattern recognition techniques permit one to define style-based automatic evaluation functions, considering *style* to be something that a set of melodies may have in common.

In this paper, we have proposed a number of diverse evaluation functions. Some of them try to describe the notes and silences in a melody statistically, comparing them with those in a training set by distance in a vector space. Other kinds of functions intend to limit the number of possibilities by introducing statistical language models inferred from the training set with the aim of giving a very low probability to melodies containing series of notes that seldom or never appear in the reference melodies. Finally, music theory-based evaluations are proposed that, by using melody segmentation algorithms or models of melodic analysis tags, try to favor those melodies with better behavior in terms of what is found for those algorithms in the training set.

The problem of combining evaluations together that are so different in such a Frankensteinian way (a term already used by Todd and Werner [1998]) can be solved by multi-objective optimization techniques, which are able to rank individuals according to their position in a fitness space relative to a geometric locus of Pareto-optimal solutions.

This research, still in an early stage, suggests that this is a flexible and powerful algorithmic composition scheme that opens the door to future studies of the influence of training data and fitness functions on system performance. The graphical interface developed is an important tool for helping to explore and refine the methodology.

Acknowledgments

The authors thank the anonymous referees for their work and especially Thomas Fiore for careful reading of the manuscript and valuable comments.

Funding

This work was supported by the Spanish Ministerio de Educación, Cultura y Deporte [FPU fellowship AP2012-0939]; and the Spanish Ministerio de Economía y Competitividad project TIMuL supported by UE FEDER funds [No. TIN2013–48152–C2–1–R].

Disclosure statement

No potential conflict of interest was reported by the authors.

Supplemental online material

Supplemental online material for this article can be accessed at http://dx.doi.org/10.1080/17459737.2016.1188171, where the reader can find some technical details, information about the data used, generated melodies, and additional information about the developed prototype and its performance.

References

Alfonseca, Manuel, Manuel Cebrián, and Alfonso Ortega. 2007. "A Simple Genetic Algorithm for Music Generation by Means of Algorithmic Information Theory." In *Proceedings of the IEEE Congress on Evolutionary Computation* (*CEC* 2007), 25–28 September 2007, Singapore, 3035–3042.

Ayesh, Aladdin, and Andrew Hugill. 2005. "Genetic Approaches for Evolving Form in Musical Composition." In *International IASTED Conference on Artificial Intelligence and Applications, part of the 23rd Multi-Conference on Applied Informatics*, 14–16 February 2005, Innsbruck, Austria, 318–321.

Baluja, Shumeet, Dean Pomerleau, and Todd Jochem. 1994. "Towards Automated Artificial Evolution for Computer Generated Images." *Connection Science* 6 (2-3): 325–354.

Biles, John A. 1994. "GenJam: A Genetic Algorithm for Generating Jazz Solos." In *Proceedings of the International Computer Music Conference*, 12–17 September 1994, Aarhus, Denmark, 131–137. http://igm.rit.edu/ ~ jabics/BilesICMC94.pdf.

Cambouropoulos, Emilios. 2001. "The Local Boundary Detection Model (LBDM) and Its Application in the Study of Expressive Timing." In *Proceedings of the International Computer Music Conference*, 17–22 September 2001, Havana, Cuba.

Cope, David. 1991. *Computers and Musical Style*. Madison, WI: A-R Editions.

de Freitas, Alan R. R., Frederico G. Guimarães, and Rogério V. Barbosa. 2012. "Ideas in Automatic Evaluation Methods for Melodies in Algorithmic Composition." In *Proceedings of the 9th Sound and Music Computing Conference (SMC 2012)*, 11–14 July 2012, Copenhagen, Denmark, 514–520.

Deb, Kalyanmoy, Samir Agrawal, Amrit Pratap, and T. Meyarivan. 2002. "A Fast and Elitist Multi-Objective Genetic Algorithm: NSGA-II." *IEEE Transactions on Evolutionary Computation* 6 (2): 182–197.

Doraisamy, Shyamala, and Stefan Rüger. 2003. "Robust Polyphonic Music Retrieval with *n*-Grams." *Journal of Intelligent Information Systems* 21 (1): 53–70.

Dostál, Martin. 2013. "Evolutionary Music Composition." In *Handbook of Optimization. From Classical to Modern Approach*, edited by Ivan Zelinka, Vaclav Snasel, and Ajith Abraham, 935–964, Volume 38 of the series *Intelligent Systems Reference Library*. Heidelberg: Springer. http://dx.doi.org/10.1007/978-3-642-30504-7_37.

Espí, David, Pedro J. Ponce de Leon, Carlos Pérez-Sancho, David Rizo, José M. Iñesta, Francisco Moreno-Seco, and Antonio Pertusa. 2007. "A Cooperative Approach to Style-Oriented Music Composition." In *Proceedings of the Twentieth International Workshop on Artificial Intelligence and Music (MUSIC-AI 2007)*, 6–12 January 2007, Hyderabad, India, 25–36. https://sao.dlsi.ua.es/repositori/grfia/pubs/186/wijcai07.pdf.

Fu, Tao-yang, Tsu-yu Wu, Chin-te Chen, Kai-chu Wu, and Ying-ping Chen. 2006. "Evolutionary Interactive Music Composition." In *Proceedings of the 8th Annual Conference on Genetic and Evolutionary Computation (GECCO'06)*, 8–12 July 2006, Seattle, WA, 1863–1864. http://dx.doi.org/10.1145/1143997.1144301.

Göksu, Hüseyin, Paul Pigg, and Vikas Dixit. 2005. "Music Composition Using Genetic Algorithms (GA) and Multilayer Perceptrons (MLP)." In *Proceedings of the First International Conference on Advances in Natural Computation (ICNC 2005)*, 27–29 August 2005, Changsha, PR China, Part III, 1242–1250.

Goldberg, David E. 1989. *Genetic Algorithms in Search, Optimization, and Machine Learning*. 1st ed. Boston, MA: Addison-Wesley Longman.

Herremans, Dorien, Kenneth Sorensen, and Darrell Conklin. 2014. "Sampling the Extrema from Statistical Models of Music Wwith Variable Neighbourhood Search." In *International Computer Music Conference (ICMC/SMC 2014)*, 22–26 October 2014, Athens, Greece, 1096–1103.

Hiller, Lejaren, and Leonard M. Isaacson. 1959. *Experimental Music: Composition with an Electronic Computer*. New York: McGraw-Hill.

Hofmann, David M. 2015. "A Genetic Programming Approach to Generating Musical Compositions." In *Proceedings of the 4th International Conference on Evolutionary and Biologically Inspired Music, Sound, Art and Design (EvoMUSART 2015)*, 8–10 April 2015, Copenhagen, Denmark, 89–100.

Holland, John H. 1975. *Adaptation in Natural and Artificial Systems*. The University of Michigan Press.

Horn, Jeffrey, Nicholas Nafpliotis, Nicholas Nafpliotis, David E. Goldberg, and David E. Goldberg. 1994. "A Niched Pareto Genetic Algorithm for Multi-Objective Optimization." In *Proceedings of the 1st IEEE International Conference on Evolutionary Computation and IEEE World Congress on Computational Intelligence*, 27–29 June 1994, Orlando, FL, 82–87. http://dx.doi.org/10.1109/ICEC.1994.350037.

Kneser, Reinhard, and Hermann Ney. 1995. "Improved Backing-Off for M-Gram Language Modeling." In *Proceedings of the 1995 International Conference on Acoustics, Speech, and Signal Processing (ICASSP '95)*, 9–12 May 1995, Detroit, MI, 181–184. http://dx.doi.org/10.1109/ICASSP.1995.479394.

Koga, Shimpei, and Makoto Fukumoto. 2014. "A Creation of Music-Like Melody by Interactive Genetic Algorithm with User's Intervention." In *Proceedings of the Human–Computer Interaction International Conference (HCI 2014) – Posters' Extended Abstracts*, 22–27 June 2014, Heraklion, Crete, Greece, Part I, 523–527. http://dx.doi.org/10.1007/978-3-319-07857-1_92.

Komatsu, Kyoko, Tomomi Yamanaka, Masami Takata, and Kazuki Joe. 2010. "A Music Composition Model with Genetic Programming." In *Proceedings of the International Conference on Parallel and Distributed Processing Techniques and Applications (PDPTA 2010)*, 12–15 July 2010, Las Vegas, NV, 686–692.

Koza, John R. 1992. *Genetic Programming: On the Programming of Computers by Means of Natural Selection*. Cambridge, MA: MIT Press.

Laine, Pauli, and Mika Kuuskankare. 1994. "Genetic Algorithms in Musical Style Oriented Generation." In *Proceedings of the 1st IEEE International Conference on Evolutionary Computation and IEEE World Congress on Computational Intelligence*, 27–29 June 1994, Orlando, FL, 858–862. http://dx.doi.org/10.1109/ICEC.1994.349942.

Liu, Chien-Hung, and Chuan-Kang Ting. 2015. "Music Pattern Mining for Chromosome Representation in Evolutionary Composition." In *IEEE Congress on Evolutionary Computation (CEC 2015)*, Sendai, Japan, 25–28 May 2015, 2145–2152.

Lo, Man Yat, and Simon M. Lucas. 2007. "N-Gram Fitness Function with a Constraint in a Musical Evolutionary System." In *Proceedings of the IEEE Congress on Evolutionary Computation (CEC 2007)*, 25–28 September 2007, Singapore, 4246–4251.

Machado, Penousal, Juan Romero, Bill Manaris, Antonino Santos, and Amílcar Cardoso. 2003. "Power to the Critics – A Framework for the Development of Artificial Art Critics." In *Proceedings of the 3rd Workshop on Creative Systems, Eighteenth International Joint Conference on Artificial Intelligence (IJCAI-03)*, 9–15 August 2003, Acapulco, Mexico, 55–64.

Miranda, Eduardo Reck, and John A. Biles, eds. 2007. *Evolutionary Computer Music*. London: Springer. http://dx.doi.org/10.1007/978-1-84628-600-1.

Moroni, Artemis, Jonatas Manzolli, Fernando Von Zuben, and Ricardo Gudwin. 1994. "Vox Populi: An Interactive Evolutionary System for Algorithmic Music Composition." *Leonardo Music Journal* 10: 49–54.

Nierhaus, Gerhard. 2008. *Algorithmic Composition: Paradigms of Automated Music Generation*. 1st ed. Vienna: Springer.

Oliwa, Tomasz Michal. 2008. "Genetic Algorithms and the abc Music Notation Language for Rock Music Composition." In *Proceedings of the Genetic and Evolutionary Computation Conference (GECCO '08)*, 12–16 July 2008, Atlanta, GA, 1603–1610.

Özcan, Ender, and Türker Erçal. 2007. "A Genetic Algorithm for Generating Improvised Music." In *Artificial Evolution, 8th International Conference, Evolution Artificielle (EA 2007)*, 29–31 October 2007, Tours, France, Revised Selected Papers, 266–277.

Pauls, Adam, and Dan Klein. 2011. "Faster and Smaller *N*-Gram Language Models." In *Proceedings of the 49th Annual Meeting of the Association for Computational Linguistics: Human Language Technologies (HLT '11)*, June 2011. http://nlp.cs.berkeley.edu/pubs/Pauls-Klein_2011_LM_paper.pdf.

Phon-Amnuaisuk, Somnuk, Edwin Hui Hean Law, and Chin Kuan Ho. 2007. "Evolving Music Generation with SOM-Fitness Genetic Programming." In *Proceedings of Applications of Evolutionary Computing, EvoWorkshops 2007: EvoCoMnet, EvoFIN, EvoIASP,EvoINTERACTION, EvoMUSART, EvoSTOC and EvoTransLog*, 11–13 April 2007, Valencia, Spain, 557–566.

Ponce de León, Pedro J., and José M. Iñesta. 2007. "A Pattern Recognition Approach for Music Style Identification Using Shallow Statistical Descriptors." *IEEE Transactions on Systems, Man and Cybernetics C* 37 (2): 248–257.

Purshouse, Robin C., Peter J. Fleming, Carlos M. Fonseca, Salvatore Greco, and Jane Shaw, eds. 2013. *Proceedings of the 7th International Conference on Evolutionary Multi-Criterion Optimization (EMO 2013)*, 19–22 March 2013, Sheffield, UK, Volume 7811 of *Lecture Notes in Computer Science*. Berlin: Springer. http://dx.doi.org/10.1007/978-3-642-37140-0.

Putnam, Jeffrey B. 1996. "A Grammar-Based Genetic Programming Technique Applied to Music Generation." In *Evolutionary Programming V: Proceedings of the 5th Annual Conference on Evolutionary Programming*, 29 February–3 March 1996, 277–286. Boston, MA: MIT Press.

Rizo, David. 2010. "Symbolic Music Comparison with Tree Data Structures." PhD thesis, Universidad de Alicante.

Rizo, David, Plácido R. Illescas, and Jose M. Iñesta. 2015. *Interactive Melodic Analysis*, Chapter 7 in *Computational Music Analysis*, 191–219. Cham, Switzerland: Springer. http://dx.doi.org/10.1007/978-3-319-25931-4_8.

Romero, Juan, and Penousal Machado, eds. 2007. *The Art of Artificial Evolution: A Handbook on Evolutionary Art and Music*. Springer series on Natural Computing. Berlin: Springer-Verlag. http://dx.doi.org/10.1007/978-3-540-72877-1.

Schaffer, David J. 1985. "Multiple Objective Optimization with Vector Evaluated Genetic Algorithms." In *Proceedings of the First International Conference on Genetic Algorithms and Their Applications*, 24–26 July 1985, Pittsburgh, PA, 93–100.

Selfridge-Field, Eleanor. 1997. *Beyond MIDI: The Handbook of Musical Codes*. Boston, MA: MIT Press.

Sheikholharam, Peyman, and Mohammad Teshnehlab. 2008. "Music Composition Using Combination of Genetic Algorithms and Recurrent Neural Networks." In *Proceedings of the 8th International Conference on Hybrid Intelligent Systems (HIS 2008)*, 10–12 September 2008, Barcelona, Spain, 350–355.

Spector, Lee, and Adam Alpern. 1995. "Induction and Recapitulation of Deep Musical Structures." In *Proceedings of the Fourteenth International Joint Conference on Artificial Intelligence (IJCAI-95), Workshop on Music and AI*, 20–25 August 1995, Montréal, Québec, Canada, 41–48.

Srinivas, Nidamarthi, and Kalyanmoy Deb. 1994. "Multi-Objective Optimization Using Nondominated Sorting in Genetic Algorithms." *Evolutionary Computation* 2 (3): 221–248.

Todd, Peter, and Eduardo R. Miranda. 2003. "Putting Some (Artificial) Life into Models of Musical Creativity." In *Musical Creativity: Current Research in Theory and Practice*, edited by Irene Deliege and Geraint Wiggins. Hove, UK: Psychology Press.

Todd, Peter M., and Gregory M. Werner. 1998. *Frankensteinian Methods for Evolutionary Music Composition*. Cambridge, MA: MIT Press/Bradford Books.

Tokui, Nao, and Hitoshi Iba. 2000. "Music Composition with Interactive Evolutionary Computation." In *Proceedings of the 3rd International Conference on Generative Art*, 215–226. http://www.generativeart.com/.

Vatolkin, Igor. 2013. "Measuring the Performance of Evolutionary Multi-Objective Feature Selection for Prediction of Musical Genres and Styles." In *Informatik 2013, 43. Jahrestagung der Gesellschaft für Informatik e.V. (GI), Informatik angepasst an Mensch, Organisation und Umwelt*, 16–20 September 2013, Koblenz, 3012–3025.

Waschka, Rodney, II. 2007. "Avoiding the Fitness Bottleneck Using Genetic Algorithms to Compose Orchestral Music." In *Proceedings of the Twentieth International Workshop on Artificial Intelligence and Music (MUSIC-AI 2007)*, 6–12 January 2007, Hyderabad, India, 25–36.

Weale, Timothy, and Jennifer Seitzer. 2003. "EVOC: A Music Generating System Using Genetic Algorithms." In *Proceedings of the Eighteenth International Joint Conference on Artificial Intelligence (IJCAI-03)*, 9–15 August 2003, Acapulco, Mexico, 1383–1384.

Wiggins, Geraint A., George Papadopoulos, Somnuk Phon-Amnuaisuk, and Andrew Tuson. 1998. "Evolutionary Methods for Musical Composition." In *Proceedings of the Workshop on Anticipation, Music and Cognition (CASYS'98)*, Liège, Belgium.

Willingham, Timothy J. 2013. "The Harmonic Implications of the Non-Harmonic Tones in the Four-Part Chorales of Johann Sebastian Bach." PhD thesis, Liberty University, Lynchburg, VA.

Xenakis, I. 1992. *Formalized Music: Thought and Mathematics in Composition*. Harmonologia series No. 6. Stuyvesant, NY: Pendragon Press.

Zhu, Hua, Shangfei Wang, and Zhen Wang. 2008. "Emotional Music Generation Using Interactive Genetic Algorithm." In *International Conference on Computer Science and Software Engineering (CSSE 2008)*, Volume 1: *Artificial Intelligence*, 12–14 December 2008, Wuhan, PR China, 345–348.

Zitzler, Eckart, Marco Laumanns, and Lothar Thiele. 2001. "SPEA2: Improving the Strength Pareto Evolutionary Algorithm for Multi-Objective Optimization." In *Proceedings of Evolutionary Methods for Design Optimization and Control with Applications to Industrial Problems (EUROGEN2001)*, 19–21 September 2001, Athens, Greece, edited by Kyriakos C. Giannakoglou, 95–100. Barcelona, Spain: International Center for Numerical Methods in Engineering (CIMNE).

Index

Note: Page numbers in *italic* type refer to figures
Page numbers in **bold** type refer to tables